NEW APPROACHES OF PROTEIN FUNCTION PREDICTION FROM PROTEIN INTERACTION NETWORKS

NEW APPROACHES OF PROTEIN FUNCTION PREDICTION FROM PROTEIN INTERACTION NETWORKS

JINGYU HOU
Deakin University, Burwood, VIC, Australia

Academic Press is an imprint of Elsevier
125 London Wall, London EC2Y 5AS, United Kingdom
525 B Street, Suite 1800, San Diego, CA 92101-4495, United States
50 Hampshire Street, 5th Floor, Cambridge, MA 02139, United States
The Boulevard, Langford Lane, Kidlington, Oxford OX5 1GB, United Kingdom

© 2017 Elsevier Ltd. All rights reserved.

No part of this publication may be reproduced or transmitted in any form or by any means, electronic or mechanical, including photocopying, recording, or any information storage and retrieval system, without permission in writing from the publisher. Details on how to seek permission, further information about the Publisher's permissions policies and our arrangements with organizations such as the Copyright Clearance Center and the Copyright Licensing Agency, can be found at our website: www.elsevier.com/permissions.

This book and the individual contributions contained in it are protected under copyright by the Publisher (other than as may be noted herein).

Notices

Knowledge and best practice in this field are constantly changing. As new research and experience broaden our understanding, changes in research methods, professional practices, or medical treatment may become necessary.

Practitioners and researchers must always rely on their own experience and knowledge in evaluating and using any information, methods, compounds, or experiments described herein. In using such information or methods they should be mindful of their own safety and the safety of others, including parties for whom they have a professional responsibility.

To the fullest extent of the law, neither the Publisher nor the authors, contributors, or editors, assume any liability for any injury and/or damage to persons or property as a matter of products liability, negligence or otherwise, or from any use or operation of any methods, products, instructions, or ideas contained in the material herein.

Library of Congress Cataloging-in-Publication Data

A catalog record for this book is available from the Library of Congress

British Library Cataloguing-in-Publication Data

A catalogue record for this book is available from the British Library

ISBN: 978-0-12-809814-1

For information on all Academic Press publications visit our website at https://www.elsevier.com/

Publisher: Glyn Jones
Acquisition Editor: Glyn Jones
Editorial Project Manager: Charlotte Cockle
Production Project Manager: Debasish Ghosh
Cover Designer: Matthew Limbert

Typeset by SPi Global, India

CONTENTS

CHAPTER 1

Introduction

Proteins are large, complex molecules of biological tissues that play major structural and functional roles in a cell. They are involved in all cell functions that make organs and the body work. Each protein has its own, specific functions. Some proteins provide structure and support for cells, while others are involved in transmitting signals to coordinate biological processes between cells, tissues and organs, or in defending the body from foreign invaders such as viruses and bacteria, or in carrying out thousands of chemical reactions to assist with biological processes in cells, such as reading the genetic information stored in DNA to form new molecules. It is obvious that proteins are the most essential and important molecules of life and that correctly annotating their functions therefore greatly helps human beings to understand various biological processes and phenomena and even to develop new methods or products to control and treat diseases such as cancers.

Until recently, there were two principal approaches to annotating protein functions: experimental and computational approaches (Eisenberg et al., 2000). The experimental approach acquires protein function knowledge from biochemical, genetic or structural experiments on an individual protein. Although all primary knowledge of protein function comes from experiments, the experimental approach is low-throughput because of the high cost of experimental and human efforts. On the other hand, advances in sequencing technologies have made the rate at which protein functions can be experimentally annotated much slower than the rate at which new sequences become available (Gabaldon and Huynen, 2004). To fill or narrow this gap, computational approaches have been developed to predict protein functions by applying computational methods to a wide variety of data generated by high-throughput technologies, such as gene and protein sequencing data, gene expression data and protein interaction data. Another important reason why the computational approach is widely accepted is that computational methods are capable of quickly predicting the functions of many unannotated proteins and of offering deep insights into protein functions from many angles. Therefore the computation approach makes it possible to flexibly combine various computational prediction results to

New Approaches of Protein Function Prediction from Protein Interaction Networks
http://dx.doi.org/10.1016/B978-0-12-809814-1.00001-7
© 2017 Elsevier Ltd.
All rights reserved.

annotate proteins more accurately. For instance, functions predicted from protein sequences or three-dimensional structures reveal the functions and their relationships within cells, while the functions predicted from protein interaction data make it possible to understand protein functions in the context of their interactions. A proper combination of predicted results from these two different domains could then make the final predicted functions closer to the real ones of the proteins. There are of course more than two computational methods available nowadays to predict protein functions from various data resources or domains. The variety of available prediction methods is a good opportunity as well as a big challenge to us on how to properly incorporate different prediction results to more accurately predict functions.

Protein function prediction relies on the definition of function. A general definition of protein function that is widely accepted was proposed by Rost et al. (2003), that is, 'function is everything that happens to or through a protein'. In real applications and research, protein functions need to be specifically defined and classified by a scheme. Since protein function is not a well-defined term, there exist several schemes that categorize protein functions. Among them Gene Ontology (GO) and Functional Catalogue (FunCat) are two commonly used schemes that are based on general biological phenomena taking place in a wide variety of organisms and eukaryotes, and they satisfy the desired properties of a scheme (Riley, 1998; Rison et al., 2000; Ouzounis et al., 2003). It is well known that proteins are the products of genes and that a protein function is encoded by a gene. So we do not distinguish gene function from protein function and consider the function of a gene as the function of its products, that is, proteins.

1.1 GENE ONTOLOGY (GO) SCHEME

The concept *ontology* originates from philosophy. It represents a formal representation of a knowledge body that consists of a set of classes or terms with relationships between them within a given domain (GO Consortium, 2016). GO represents three gene and protein function domains at the highest level: biological process, molecular function and cellular component. Each domain is hierarchically organized with functional labels/terms and their relations. All these three domains join together to form an overall structure of GO, which is modelled as a directed acyclic graph (DAG). Each node of this DAG corresponds to a functional label and each directed edge corresponds to a '*is a*' (I) or '*part of*' (P) or '*regulates*' (R) relationship. Fig. 1.1 illustrates this GO graph model.

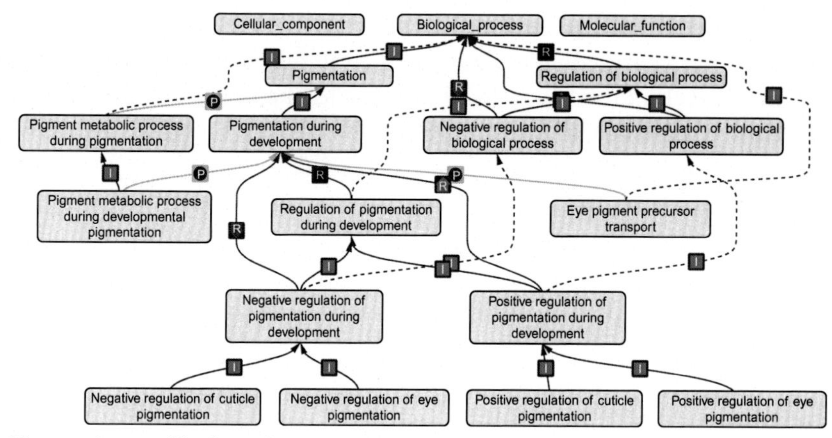

Fig. 1.1 A part of 'biological process' domain in GO. *(Adapted from GO Consortium, 2016. Ontology structure. http://geneontology.org/page/ontology-structure (accessed 11.03.16)).*

Within the GO hierarchical structure, 'child' terms are more specialized than their 'parent' terms, and a 'child' term may have more than one 'parent' term. Every term is identified by an ID which consists of a term name and a digital number prefixed by 'GO:', for example, the term 'glucose transport (GO: 0015758)'. The numerical portion of the ID, for example, 0015758, is only a reference to the source of the information and can be used to trace who added the term. It is not related to the position of the term in the ontologies.

GO is a species-independent integrated resource of function information for genes from over 460,000 species covering plants, animals, and the microbial world (GO Consortium, 2015). This characteristic ensures the wide applicability of the scheme, including gene product functional classification and prediction.

1.2 FUNCAT SCHEME

FunCat was developed by Munich Information Center for Protein Sequences (MIPS). It is a hierarchically structured, organism-independent, flexible and scalable controlled classification system enabling the functional description of protein from any organism (Ruepp et al., 2004).

The FunCat annotation scheme consists of 28 main categories (branches) of protein functions such as metabolism and protein activity regulation. Each category (branch) is represented in the FunCat hierarchical structure as a tree, and there are a total of 1307 categories (Ruepp et al., 2004). Each category in a certain level of the hierarchy is assigned a unique two-digit number. The upward context or the ancestors of a category is indicated by the prefix of the preceding nodes (two-digit numbers separated by dots) in the upper levels of the hierarchy. For example, the function '*biosynthesis of glutamate*' is presented as '*01.01.03.02.01*' which is in the fifth level of the hierarchy. Its ancestor functions are *01.01.03.02* (*metabolism of glutamate*), *01.01.03* (*assimilation of ammonia, metabolism of the glutamate group*), *01.01* (*amino acid metabolism*), and the highest level *01* (*metabolism*). Fig. 1.2 shows a part of FunCat hierarchy.

Rison et al. (2000) conducted a quantitative comparison of the existing six schemes, and the result showed that FunCat had the best coverage and generality. Different from GO which is not strictly hierarchical, FunCat has a simple and intuitive hierarchical structure which makes it possible to easily browse through the main categories down to the specific level of functions. Owing to these characteristics, FunCat has proved to be a useful tool for many bioinformatics research projects and applications and is one of the most prevalent annotation schemes.

With the well-defined functional annotation schemes such as GO and FunCat, various computational approaches have been proposed for protein function prediction. Given the wide range of available computational techniques, we take the similar way that Pandey et al. (2006) used to generally classify these computational approaches based on the main types of biological data:

(1) Amino acid sequences
(2) Protein structure
(3) Genome sequences
(4) Phylogenetic data
(5) Microarray expression data

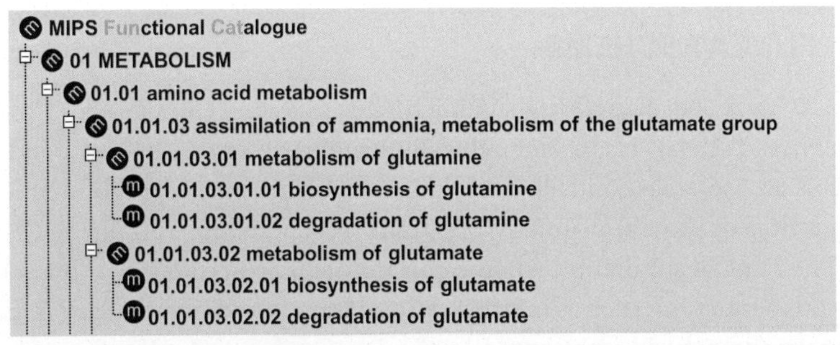

Fig. 1.2 A part of FunCat scheme. *(Adapted from MIPS FunCat database at http://mips.helmholtz-muenchen.de/funcatDB/).*

(6) Protein interaction networks
(7) Biomedical literature
(8) Combination of multiple data types

1.3 APPROACHES BASED ON AMINO ACID SEQUENCES

Proteins are made up of hundreds or thousands of smaller units called amino acids, which are attached to one another to form long chains. There are 20 different types of amino acids that can be combined to make a protein. The sequence of amino acids in chains is the most fundamental form of information about a protein because it determines each protein's unique three-dimensional structure, subcellular localization and its specific function. So the rationale behind this approach is that two proteins with similar sequence or structure could evolve from a common ancestor and thus have similar functions.

The representative sequence similarity system is the Basic Local Alignment Search Tool (BLAST; Altschul et al., 1990), which was enhanced into PSI-BLAST later (Altschul et al., 1997). The idea of using sequence similarity to predict protein function is to use a sequence similarity system, such as BLAST, to search standard protein sequence databases like SWISS-PROT (Boeckmann et al., 2003) to find proteins that have a higher sequence similarity ranking (i.e., the sequence similarity score is greater than a predefined threshold) with the unannotated protein. Then the functions of these most homologous proteins are transferred to the unannotated protein as the predicted functions. Based on this idea, improved approaches were proposed to enhance the prediction effectiveness, such as subsequence-based approaches and feature-based approaches which extract segments or subsequences as the features of a protein sequence and construct models to map these features to protein functions. The prediction is then based on these models, rather than the whole sequence similarity (Pandey et al., 2006).

Although the applications of this approach have produced some promising prediction results, other studies (Gerlt and Babbitt, 2000; Devos and Valencia, 2000; Whisstock and Lesk, 2003; Kuang et al., 2005a,b; Rangwala and Karypis, 2005) show that sequence similarity is more likely correlated with the structural than the functional similarity of proteins. These studies also highlight the limitations of this prediction approach. However, this approach suggests a new way of combining sequence similarity with other information resources or models to more effectively predict functions (Sandhan et al., 2015).

1.4 APPROACHES BASED ON PROTEIN STRUCTURE

Proteins are biological polymers composed of amino acids. Amino acids are linked together by peptide bonds to form a sequence called polypeptide chain. One or more polypeptide chains are twisted into a three-dimensional shape to form a protein that is able to perform its biological function. There are four distinct levels of protein structure: primary structure, secondary structure, tertiary structure and quaternary structure. Because a protein's biological function is related to its structure, the idea of protein structure-based function prediction is to establish the relationship between protein structure and function and to exploit this relationship to predict functions from structures.

However, many studies (Martin et al., 1998; Hegyi and Gerstein, 1999; Orengo et al., 1999; Thornton et al., 1999) demonstrated that the correlation between structure and function is not strong enough to enable the direct function prediction from protein structures. Instead protein structures more likely correlate with lower-level functional features of proteins. Therefore a possible path suggested was to convert a protein's structure into lower-level functional features based on which function prediction could be conducted more robustly. This idea led to some new function prediction approaches being proposed. These approaches can be generally classified into four categories (Pandey et al., 2006): *Similarity-Based Approach* which predicts functions by identifying the protein that has the most structure similarity with the unannotated protein and assigns its functions to the unannotated protein; *Motif-Based Approach* which creates a mapping between protein functions and motifs (substructures) and uses this mapping to predict the functions of the unannotated proteins; *Surface-Based Approach* which identifies the features of a protein's continuous surface and utilizes these features to predict the functions of a protein and lastly the *Learning-Based Approach* which employs effective classification methods to identify the most appropriate functional class for a protein from its most relevant structural features.

1.5 APPROACHES BASED ON GENOME SEQUENCES

The availability of genomic sequences in many databases such as GeneBank (Benson et al., 2004) makes bioinformatics researchers seek possible ways of predicting protein function by exploiting the information in genomic sequences. The primary approach is to use sequence search systems, such as PSI-BLAST (Altschul et al., 1997), to search for homologous sequences from genomic sequence databases and obtain functional information from a large number of proteins and organisms. Then the function of a query protein is predicted as the function of a protein that is most similar to the query protein in terms of genomic sequence similarity. This is a popular prediction approach based on protein and genomic sequences.

Most genomic sequence-based approaches intend to reveal functional associations between genes or proteins or investigate evolution mechanisms of genes, rather than directly predicting functions for individual proteins (Marcotte, 2000; Koonin and Galperin, 2002). However, the functional associations between proteins could be further used to predict interactions between proteins, and in turn to predict functions of unannotated proteins. One representative approach (also known as gene neighbourhood-based approach) in this category is to infer functional associations between genes and their corresponding proteins based on the observation that two or more proteins, whose corresponding genes are 'close' to each other on a genome, are functionally related (Dandekar et al., 1998; Overbeek et al., 1999). Another approach is based on gene fusion (Marcotte et al., 1999; Yanai et al., 2001). The idea of this approach is that if two separate genes in one genome are fused/merged as a single gene in another genome, then these genes are expected to be functional related.

1.6 APPROACHES BASED ON PHYLOGENETIC DATA

The approaches in this category exploit evolution-based data for protein function prediction which rely on a fact in the evolution theory that changes in the physiologies of different organisms are driven by the changes at the cellular level, including the adoption and surrender of functions by proteins because of the changes in genes. Therefore evolution data provide another clue to understanding and predicting protein function. Generally the common forms of evolution data are phylogenetic profiles and phylogenetic trees. The phylogenetic profile of a protein is primarily represented as a vector whose length is the number of available genomes and where the element value in the *i*th position of the vector is either 1, if the *i*th genome contains a homologues of the corresponding gene, or 0 otherwise. There are some variations of these vectors whose element values are real numbers instead of 1s or 0s to reflect the similarity between the original gene and the best match in the genome. Phylogenetic tree is a more extensive representation of evolution knowledge (Baldauf, 2003). It has to be constructed by using tools with various data mining and probabilistic methods, such as PHYLIP (Felsenstein, 1989) and PAML (Yang, 1997). The leaves of the tree correspond to the organisms used to construct the tree, an internal node denotes the hypothetical last common ancestor of all its descendants and a tree branch indicates the '*has evolved from*' relationship.

The function prediction approaches using phylogenetic profiles are based on the assumption that proteins with similar phylogenetic profiles are functionally related (Pellegrini et al., 1999; Liberles et al., 2002; Wu et al., 2003; Date and Marcotte, 2005). The methods in this category therefore are primarily comparative in nature and define different ways to measure the similarity between two protein profiles and predict functions from the annotated protein that is most similar to the unannotated protein. The prediction approaches that are based on phylogenetic trees, which have a richer knowledge of genetic evolution, apply various data mining and machine learning methods to extract functional relationships between proteins for function prediction (Eisen, 1998; Doerks et al., 1998; Sjolander, 2004; Engelhardt et al., 2005). The effectiveness of this approach largely relies on the proper and accurate construction of the phylogenetic trees, which is also a challenge this approach has to face. There are also many hybrid approaches that combine evolution knowledge from both phylogenetic profiles and trees to predict protein functions (Vert, 2002; Narra and Liao, 2005).

1.7 APPROACHES BASED ON MICROARRAY EXPRESSION DATA

Microarray expression data shows simultaneous activities of thousands of genes under a certain condition. It therefore has great promise for determining the function and functional associations of proteins which are the products of genes. On the other hand, the matrix format of microarray expression data makes it easy to process by computer algorithms. Because of these advantages of microarray expression data, a lot of computational approaches have been developed to exploit the expression data, also known as gene expression profiles, to predict protein functions. These approaches can be classified in three general categories (Pandey et al., 2006): clustering-based, classification-based and temporal analysis-based.

The clustering-based approach is based on the hypothesis that genes with similar expression files are functionally similar (Eisen et al., 1998; Ben-Dor et al., 1999; Ng et al., 2004; Swift et al., 2004; Madeira and Oliveira, 2004;

Liu et al., 2004; Banerjee et al., 2005). Approaches in this category define gene similarities on the basis of gene expression profiles and cluster similar genes into groups. The functions of an unannotated protein are predicted by the most dominate functions of the group within which the unannotated protein is located.

The classification-based approach uses various classifiers, such as neural network, SVM and naïve Bayes classifier in data mining, to build various types of models that map gene expression profiles to functions and to predict functions of unannotated proteins from these models (Brown et al., 2000; Mateos et al., 2002; Ng et al., 2003; Zhang et al., 2004; Kuramochi and Karypis, 2005).

The temporal analysis-based approach uses temporal gene expression profiles that are measured at different time instances to analyse the activities of genes, derive features from the temporal data, and then predict functions of unannotated proteins using classification techniques (Hvidsten et al., 2001; Laegreid et al., 2003; Gui and Li, 2003; Moller-Levet et al., 2003; Deng and Ali, 2004; Bar-Joseph, 2004; Jiang et al., 2004; Ernst et al., 2005; Heard et al., 2005).

1.8 APPROACHES BASED ON PROTEIN INTERACTION NETWORKS

It has been known that a protein does not perform its function in isolation. Instead it usually interacts with other proteins to perform a certain function. Thanks to novel high-throughput technologies of protein interaction identification (Aebersold and Mann, 2003; Field, 2005), large-scale databases, such as MIPS (Mewes et al., 2002), DIP (Salwinski et al., 2004) and BioGRID (Chatr-Aryamontri et al., 2015) on protein interactions across human and most model species have been created. These protein interactions can be modelled as an undirected network with nodes representing proteins and edges representing the detected pairwise interactions. Protein interaction networks provide a context of a protein within which it performs its function as a part of the biological processes in an organism. Because of the rich information it carries, protein interaction network is a promising and major data resource for protein function prediction, based on which a lot of approaches have been developed. The current approaches in this category can be generally classified in the following subcategories: direct, module-based and integration approaches.

The direct approaches attempt to predict functions by using the topological structure information of a protein interaction network. The primary form of this category is to use the local topological information, that is, the interaction neighbours of an unannotated protein, to predict functions (Schwikowski et al., 2000; Hishigaki et al., 2001; Samanta and Liang, 2003; Brun et al., 2003; Chua et al., 2006). To overcome the limitations of using local topological information for function prediction, such as an insufficient number of neighbours or the neighbour proteins are unannotated, other approaches are proposed that try to take into account the full topology of the network and optimize an objective function (directly or indirectly) by properly assigning functions to the unannotated proteins in the network (Deng et al., 2003; Vazquez et al., 2003; Karaoz et al., 2004; Nabieva et al., 2005).

The module-based approaches predict function first by identifying coherent groups or modules of proteins from the protein interaction network and then assigning the dominant functions to all unannotated proteins in each group or module. The approaches in this category differ mainly in the techniques they use to identify modules (Bader and Hogue, 2003; Przulj et al., 2004; King et al., 2004; Sharan et al., 2005; Arnau et al., 2005; Altaf-Ul-Amin et al., 2006; Trivodaliev et al., 2014).

The integration approaches combine protein interaction data with diverse additional sources to predict protein functions, making protein interaction networks as a solid platform that integrates the interplay of different information sources for understanding biological processes and predicting protein functions as well. The spectrum of this approach varies depending on what and how other information is integrated with protein interaction networks, and accordingly how to define and identify the predicted functions for the unannotated proteins. A typical approach is to integrate protein interaction networks with microarray expression profiles, protein sequences or/and protein complexes for function prediction (Zien et al., 2000; Jansen et al., 2002; Tornow and Mewes, 2003; Simonis et al., 2004; Luscombe et al., 2004; Balazsi et al., 2005; de Lichtenberg et al., 2005; Wachi et al., 2005; Lan et al., 2013; Peng et al., 2014; Wu et al., 2014). Currently, there are more and more approaches that integrate semantic information, such as GO structural information and semantic similarities, with protein interaction networks to predict functions (Jiang et al., 2008; Cho et al., 2008; Pandey et al., 2009; Hu et al., 2010; Cozzetto et al., 2013; Piovesan et al., 2015).

1.9 APPROACHES BASED ON BIOMEDICAL LITERATURE

Huge repositories, such as MEDLINE/PubMed (https://www.nlm.nih.gov/bsd/pmresources.html), of biology and medicine knowledge in the form of papers, books, reports and theses contain a huge amount of useful information that can be mined for various biomedical research, including protein function prediction. Protein function prediction from biomedical literature mining aims to extract relevant and valid biological information from vast repositories and uncover those functions that have not been reported in the literature.

The biomedical literature-based approaches can be classified into three categories according to the underlying techniques: information retrieval (IR)-based, data mining-based and natural language processing (NLP)-based (Pandey et al., 2006). The information retrieval-based approach uses information retrieval methods to find documents in repositories that are relevant to the query (i.e., a gene or a protein) and then rank them to get the most relevant results (Tamames et al., 1998; Couto et al., 2003; Rubinstein and Simon, 2005).

The text mining-based approach uses data analysis techniques, such as clustering and classification, to analyse text data obtained from the repositories and extract semantic-related knowledge for function prediction (Renner and Aszodi, 2000; Raychaudhari et al., 2002; Keck and Wetter, 2003; Stoica and Hearst, 2006).

The natural language processing-based approach is to model and analyse the natural language with text mining techniques to better understand queries and retrieve semantic-related knowledge from text in order to predict functions (Koike et al., 2005; Chiang and Yu, 2005).

1.10 APPROACHES BASED ON THE COMBINATION OF MULTIPLE DATA TYPES

The approaches in this category aim to combine different types of data sources, such as gene mutant network (Rung et al., 2002), gene in silico network (Pilpel et al., 2001), protein complexes and so on, to more accurately and effectively predict protein functions. The motivation is that the combination can provide a global picture of the biological phenomena that a set of genes/proteins is involved in, improve the data set quality and validate the predictions across a set of data types (Kemmeren and Holstege, 2003). The approaches in this category are classified into two groups based on the ways of dealing with the variety of data sources.

The first group of this approach category consists of those approaches that try to transform original data types into a single common format using proper data preprocessing techniques. Based on the transformed data sources in a common format, an existing technique is applied to the whole data set to predict functions, or different techniques are applied to different data types to derive different predictions from which final consensus predictions are made (Schlitt et al., 2003; Strong et al., 2003; Schomburg et al., 2004; Chen and Xu, 2004; Kemmeren et al., 2005; Peng et al., 2014; Yu et al., 2015).

Approaches in the second group try to retain the original formats of different data types and apply separate techniques to derive predictions individually, based on which the final predictions are made by combining individual prediction results to derive a consensus set of predictions, or by modelling the dependence between the individual prediction results to make final predictions using machine learning techniques (Xie et al., 2002; Pavlidis et al., 2002; Troyanskaya et al., 2003; Pal and Eisenberg, 2005; Tsuda et al., 2005; Barutcuoglu et al., 2006; Cozzetto et al., 2013; Lan et al., 2013; Wu et al., 2014).

The approaches in the second group are more popular as they can avoid the information loss incurred in transforming all data types into a common format which is performed by the first group approaches.

The preceding brief introductions to the approaches of protein function prediction show that this research area is very active because of the importance of proteins in various biological processes. Compared with other approaches, protein interaction-based approaches attract much more attention from researchers, generate satisfactory prediction results in many cases and have become a prevalent tool for functional annotation. This is because the protein interaction information more reasonably reflects the context within which proteins perform their biological functions and provides a global picture of interrelationship between proteins. This makes it possible to predict protein function from a systematic point of view, rather than from a separate and individual angle. Furthermore, protein interaction networks provide a flexible and solid platform that enables the integration of various data sources (such as gene expression profiles, GO, protein sequences and so on) and techniques (such as data mining, machine learning, graph theory, optimization theory and probabilistic and statistic methods) for function prediction and other research such as protein complex and functional module detection.

However, there are several issues associated with protein interaction-based approaches. The most important one is the large amount of noise information contained in high-throughput interaction data because of the errors and limitations of experimental approaches used for collecting data or the limitations of computational methods used for predicting protein interactions (Salwinski and Eisenberg, 2003; Deng et al., 2003). Another important issue is that protein interaction data/network lacks dynamic information of protein interactions, because for some functions the relevant interactions may occur only under specific conditions at a specific point in time (Sharan et al., 2007). Meanwhile protein interaction data does not reveal the functional strength of interactions. These issues make the protein function prediction an open and challenging research area, although numerous approaches and techniques have been proposed.

This book does not intend to be an encyclopaedia of protein interaction-based function prediction methods. Instead it introduces some new and innovative approaches, addressing the preceding important issues in this area. Specifically, Chapter 2 introduces a new method for assessing and selecting reliable interactions from original interaction data sets to improve data quality and reduce the impact of noise data on the effectiveness and accuracy of prediction. Chapter 3 presents a new clustering-based prediction method for layered function prediction, and a dynamic clustering-based approach for predicting functions, which reflects the dynamic nature of protein interaction in the clustering process. Along with these methods, new semantic similarities between proteins are defined accordingly. Chapter 4 provides an iterative approach for function prediction, which exploits the dynamic mutual functional relationship between proteins in their interactions. Chapter 5 presents a mathematic model and the corresponding algorithms to incorporate functional aggregation into the dynamic function prediction. The functional aggregation reveals the functional relationship and strength between proteins in the prediction process. Chapter 6 introduces an innovative approach of reducing the impact of noise data on function prediction by dynamically selecting a proper prediction domain from a protein interaction network. Chapter 7 introduces a new concept of functional connectivity to measure the functional importance of proteins within an interaction network and presents a novel function prediction method that incorporates the protein functional connectivity to improve prediction results. Finally, Chapter 8 summarizes the main contents of this book and discusses possible future research directions in this area.

REFERENCES

Aebersold, R., Mann, M., 2003. Mass spectrometry-based proteomics. Nature 422, 198–207.

Altaf-Ul-Amin, M., Shinbo, Y., Mihara, K., Kurokawa, K., Kanaya, S., 2006. Development and implementation of an algorithm for detection of protein complexes in large interaction networks. BMC Bioinf. 7, 207.

Altschul, S.F., Gish, W., Miller, W., Myers, E.W., Lipman, D.J., 1990. Basic local alignment search tool. J. Mol. Biol. 215 (3), 403–410.
Altschul, S.F., Madden, T.L., Schffer, A.A., Zhang, J., Zhang, Z., Miller, W., Lipman, D.J., 1997. Gapped BLAST and PSI-BLAST: a new generation of protein database search programs. Nucleic Acids Res. 25 (17), 3389–3402.
Arnau, V., Mars, S., Marin, I., 2005. Iterative cluster analysis of protein interaction data. Bioinformatics 21, 364–378.
Bader, G.D., Hogue, C.W., 2003. An automated method for finding molecular complexes in large protein interaction networks. BMC Bioinf. 4, 2.
Balazsi, G., Barabasi, A.L., Oltvai, Z.N., 2005. Topological units of environmental signal processing in the transcriptional regulatory network of *Escherichia coli*. Proc. Natl. Acad. Sci. U. S. A. 102, 7841–7846.
Baldauf, S.L., 2003. Phylogeny for the faint of heart: a tutorial. Trends Genet. 19 (6), 347–351.
Banerjee, A., Krumpelman, C., Ghosh, J., Basu, S., Mooney, R.J., 2005. Model-based overlapping clustering. In: Proceedings of the 11th ACM SIGKDD International Conference on Knowledge Discovery and Data Mining (KDD), pp. 532–537.
Bar-Joseph, Z., 2004. Analyzing time series gene expression data. Bioinformatics 20 (16), 2493–2503.
Barutcuoglu, Z., Schapire, R.E., Troyanskaya, O.G., 2006. Hierarchical multi-label prediction of gene function. Bioinformatics 22 (7), 830–836.
Ben-Dor, A., Shamir, R., Yakhini, Z., 1999. Clustering gene expression patterns. J. Comput. Biol. 6 (3-4), 281–297.
Benson, D.A., Karsch-Mizrachi, I., Lipman, D.J., Ostell, J., Wheeler, D.L., 2004. Genbank: update. Nucleic Acids Res. 32 (Database issue), D23–D26.
Boeckmann, B., Bairoch, A., Apweiler, R., Blatter, M.-C., Estreicher, A., Gasteiger, E., Martin, M.J., Michoud, K., O'Donovan, C., Phan, I., Pilbout, S., Schneider, M., 2003. The SWISS-PROT protein knowledgebase and its supplement TrEMBL in 2003. Nucleic Acids Res. 31 (1), 365–370.
Brown, M.P., Grundy, W.N., Lin, D., Cristiniani, N., Sugnet, C.W., Furey, T.S., Ares Jr., M., Haussler, D., 2000. Knowledge based analysis of microarray gene expression data by using support vector machines. Proc. Natl. Acad. Sci. U. S. A. 97 (1), 262–267.
Brun, C., Chevenet, F., Martin, D., Wojcik, J., Guenoche, A., Jacq, B., 2003. Functional classification of proteins for the prediction of cellular function from a protein-protein interaction network. Genome Biol. 5 (1), R6.
Chatr-Aryamontri, A., Breitkreutz, B.J., Oughtred, R., Boucher, L., Heinicke, S., Chen, D., Stark, C., Breitkreutz, A., Kolas, N., O'Donnell, L., Reguly, T., Nixon, J., Ramage, L., Winter, A., Sellam, A., Chang, C., Hirschman, J., Theesfeld, C., Rust, J., Livstone, M.S., Dolinski, K., Tyers, M., 2015. The BioGRID interaction database: 2015 update. Nucleic Acids Res. 43 (Database issue), D470–D478.
Chen, Y., Xu, D., 2004. Global protein function annotation through mining genome-scale data in yeast *Saccharomyces cerevisiae*. Nucleic Acids Res. 32 (21), 6414–6424.
Chiang, J.-H., Yu, H.-C., 2005. Literature extraction of protein functions using sentence pattern mining. IEEE Trans. Knowl. Data Eng. 17 (8), 1088–1098.
Cho, Y.-R., Shi, L., Ramanathan, M., Zhang, A., 2008. A probabilistic framework to predict protein function from interaction data integrated with semantic knowledge. BMC Bioinf. 9, 382.
Chua, H.N., Sung, W.-K., Wong, L., 2006. Exploiting indirect neighbours and topological weight to predict protein function from protein-protein interactions. Bioinformatics 22 (13), 1623–1630.

Couto, F., Silva, M., Coutinho, P., 2003. ProFAL: protein functional annotation through literature. In: Proceedings of the 8th Conference on Software Engineering and Databases (JISBD), pp. 747–756.

Cozzetto, D., Buchan, D.W.A., Bryson, K., Jones, D.T., 2013. Protein function prediction by massive integration of evolutionary analyses and multiple data sources. BMC Bioinf. 14 (Suppl. 3), S1.

Dandekar, T., Snel, B., Huynen, M., Bork, P., 1998. Conservation of gene order: a fingerprint of proteins that physically interact. Trends Biochem. Sci. 23 (9), 324–328.

Date, S.V., Marcotte, E.M., 2005. Protein function prediction using the Protein Link EXplorer (PLEX). Bioinformatics 21 (10), 2558–2559.

de Lichtenberg, U., Jensen, L.J., Brunak, S., Bork, P., 2005. Dynamic complex formation during the yeast cell cycle. Science 307, 724–727.

Deng, X., Ali, H.H., 2004. A hidden Markov model for gene function prediction from sequential expression data. In: Proceedings of the Computational Systems Bioinformatics, pp. 670–671.

Deng, M., Zhang, K., Mehta, S., Chen, T., Sun, F., 2003. Prediction of protein function using protein–protein interaction data. J. Comput. Biol. 10 (6), 947–960.

Devos, D., Valencia, A., 2000. Practical limits of function prediction. Proteins 41 (1), 98–107.

Doerks, T., Bairoch, A., Bork, P., 1998. Protein annotation: detective work for function prediction. Trends Genet. 14 (6), 248–250.

Eisen, J.A., 1998. Phylogenomics: improving functional predictions for uncharacterized genes by evolutionary analysis. Genome Res. 8 (3), 163–167.

Eisen, M.B., Spellman, P.T., Browndagger, P.O., Botstein, D., 1998. Cluster analysis and display of genome-wide expression patterns. Proc. Natl. Acad. Sci. U. S. A. 95 (25), 14863–14868.

Eisenberg, D., Marcotte, E.M., Xenarios, I., Yeates, T.O., 2000. Protein function in the post-genomic era. Nature 405, 823–826.

Engelhardt, B.E., Jordan, M.I., Muratore, K.E., Brenner, S.E., 2005. Protein molecular function prediction by Bayesian phylogenomics. PLoS Comput. Biol. 1 (5), e45.

Ernst, J., Nau, G.J., Bar-Joseph, Z., 2005. Clustering short time series gene expression data. Bioinformatics 21 (Suppl. 1), i159–i168.

Felsenstein, J., 1989. PHYLIP—phylogeny inference package (version 3.2). Cladistics 5, 164–166.

Fields, S., 2005. High-throughput two-hybrid analysis. The promise and the peril. FEBS J. 272 (21), 5391–5399.

Gabaldon, T., Huynen, M.A., 2004. Prediction of protein function and pathways in the genome era. Cell. Mol. Life Sci. 61 (7–8), 930–944. http://dx.doi.org/10.1007/s00018-003-3387-y PMID 15095013.

Gerlt, J.A., Babbitt, P.C., 2000. Can sequence determine function? Genome Biol. 1 (5). REVIEWS0005.

GO Consortium, 2015. Gene Ontology Consortium: going forward. Nucleic Acids Res. 43 (Database issue), D1049–D1056.

GO Consortium, 2016. Ontology structure. http://geneontology.org/page/ontology-structure (accessed 11.03.16).

Gui, J., Li, H., 2003. Mixture functional discriminant analysis for gene function classification based on time course gene expression data. In: Proceedings of the Joint Statistical Meeting (Biometrics Section).

Heard, N.A., Holmes, C.C., Stephens, D.A., Hand, D.J., Dimopoulos, G., 2005. Bayesian coclustering of anopheles gene expression time series: study of immune defense response

to multiple experimental challenges. Proc. Natl. Acad. Sci. U. S. A. 102 (47), 16939–16944.
Hegyi, H., Gerstein, M., 1999. The relationship between protein structure and function: a comprehensive survey with application to the yeast genome. J. Mol. Biol. 288 (1), 147–164.
Hishigaki, H., Nakai, K., Ono, T., Tanigami, A., Takagi, T., 2001. Assessment of prediction accuracy of protein function from protein–protein interaction data. Yeast 18 (6), 523–531.
Hu, P., Jiang, H., Emili, A., 2010. Predicting protein functions by relaxation labelling protein interaction network. BMC Bioinf. 11 (Suppl. 1), S64.
Hvidsten, T., Komorowski, J., Sandvik, A., Laegreid, A., 2001. Predicting gene function from gene expressions and ontologies. In: Proceedings of the Pacific Symposium on Biocomputing (PSB), pp. 299–310.
Jansen, R., Greenbaum, D., Gerstein, M., 2002. Relating whole-genome expression data with protein–protein interactions. Genome Res. 12, 37–46.
Jiang, D., Pei, J., Ramanathan, M., Tang, C., Zhang, A., 2004. Mining coherent gene clusters from gene-sample-time microarray data. In: Proceedings of the Tenth ACM SIGKDD International Conference on Knowledge Discovery and Data Mining (KDD), pp. 430–439.
Jiang, X., Nariai, N., Steffen, M., Kasif, S., Kolaczyk, E.D., 2008. Integration of relational and hierarchical network information for protein function prediction. BMC Bioinf. 9, 350.
Karaoz, U., Murali, T.M., Letovsky, S., Zheng, Y., Ding, C., Cantor, C.R., Kasif, S., 2004. Whole-genome annotation by using evidence integration in functional-linkage networks. Proc. Natl. Acad. Sci. U. S. A. 101 (9), 2888–2893.
Keck, H.-P., Wetter, T., 2003. Functional classification of proteins using a nearest neighbour algorithm. In Silico Biol. 3 (3), 265–275.
Kemmeren, P., Holstege, F., 2003. Integrating functional genomics data. Biochem. Soc. Trans. 31 (6), 1484–1487.
Kemmeren, P., Kockelkorn, T.T.J.P., Bijma, T., Donders, R., Holstege, F.C.P., 2005. Predicting gene function through systematic analysis and quality assessment of high-throughput data. Bioinformatics 21 (8), 1644–1652.
King, A.D., Przulj, N., Jurisica, I., 2004. Protein complex prediction via cost-based clustering. Bioinformatics 20, 3013–3020.
Koike, A., Niwa, Y., Takagi, T., 2005. Automatic extraction of gene/protein biological functions from biomedical text. Bioinformatics 21 (7), 1227–1236.
Koonin, E.V., Galperin, M.Y., 2002. Sequence–Evolution–Function: Computational Approaches in Comparative Genomics. Springer, Dordrecht.
Kuang, R., Ie, E., Wang, K., Wang, K., Siddiqi, M., Freund, Y., Leslie, C., 2005a. Profile-based string kernels for remote homology detection and motif extraction. J. Bioinform. Comput. Biol. 3 (3), 527–550.
Kuang, R., Weston, J., Noble, W.S., Leslie, C., 2005b. Motif-based protein ranking by network propagation. Bioinformatics 21 (19), 3711–3718.
Kuramochi, M., Karypis, G., 2005. Gene classification using expression profiles: a feasibility study. Int. J. Artif. Intell. Tools 14 (4), 641–660.
Laegreid, A., Hvidsten, T.R., Midelfart, H., Komorowski, J., Sandvik, A.K., 2003. Predicting gene ontology biological process from temporal gene expression patterns. Genome Res. 13 (5), 965–979.
Lan, L., Djuric, N., Guo, Y., Vucetic, S., 2013. MS-kNN: protein function prediction by integrating multiple data sources. BMC Bioinf. 14 (Suppl. 3), S8.
Liberles, D.A., Thorn, A., Von Heijne, G., Elofsson, A., 2002. The use of phylogenetic profiles for gene predictions. Curr. Genomics 3 (3), 131–137.

Liu, J., Wang, W., Yang, J., 2004. Gene ontology friendly biclustering of expression profiles. In: Proceedings of the IEEE Computational Systems Bioinformatics Conference (CSB), pp. 436–447.

Luscombe, N.M., Babu, M.M., Yu, H., Snyder, M., Teichmann, S.A., Gerstein, M., 2004. Genomic analysis of regulatory network dynamics reveals large topological changes. Nature 431, 308–312.

Madeira, S.C., Oliveira, A.L., 2004. Biclustering algorithms for biological data analysis: a survey. IEEE/ACM Trans. Comput. Biol. Bioinf. 1 (1), 24–45.

Marcotte, E.M., 2000. Computational genetics: finding protein function by nonhomology methods. Curr. Opin. Struct. Biol. 10 (3), 359–365.

Marcotte, E.M., Pellegrini, M., Ng, H.-L., Rice, D.W., Yeates, T.O., Eisenberg, D., 1999. Detecting protein function and protein–protein interactions from genome sequences. Science 285 (5428), 751–753.

Martin, A., Orengo, C., Hutchinson, E., Jones, S., Karmirantzou, M., Laskowski, R., Mitchell, J., Taroni, C., Thornton, J., 1998. Protein folds and functions. Structure 6 (7), 875–884.

Mateos, A., Dopazo, J., Jansen, R., Tu, Y., Gerstein, M., Stolovitzky, G., 2002. Systematic learning of gene functional classes from DNA array expression data by using multilayer perceptrons. Genome Res. 12 (11), 1703–1715.

Mewes, H.W., Frishman, D., Guldener, U., Mannhaupt, G., Mayer, K., Mokrejs, M., Morgenstern, B., Munsterkotter, M., Rudd, S., Weil, B., 2002. MIPS: a database for genomes and protein sequences. Nucleic Acids Res. 30 (1), 31–34.

Moller-Levet, C.S., Cho, K.-H., Yin, H., Wolkenhauer, O., 2003. Clustering of gene expression time-series data: Technical Report. Department of Computer Science, University of Rostock, Germany.

Nabieva, E., Jim, K., Agarwal, A., Chazelle, B., Singh, M., 2005. Whole-proteome prediction of protein function via graph-theoretic analysis of interaction maps. Bioinformatics 21 (Suppl. 1), i1–i9.

Narra, K., Liao, L., 2005. Use of extended phylogenetic profiles with E-values and support vector machines for protein family classification. Int. J. Comput. Inform. Sci. 6(1).

Ng, S.-K., Tan, S.-H., Sundararajan, V., 2003. On combining multiple microarray studies for improved functional classification by whole-dataset feature selection. Genome Inform. 14, 44–53.

Ng, S.-K., Zhu, Z., Ong, Y.-S., 2004. Whole-genome functional classification of genes by latent semantic analysis on microarray data. In: Proceedings of the Second Asia-Pacific Conference on Bioinformatics, pp. 123–129.

Orengo, C.A., Todd, A.E., Thornton, J.M., 1999. From protein structure to function. Curr. Opin. Struct. Biol. 9 (3), 374–382.

Ouzounis, C.A., Coulson, R.M., Enright, A.J., Kunin, V., Pereira-Leal, J.B., 2003. Classification schemes for protein structure and function. Nat. Rev. Genet. 4 (7), 508–519.

Overbeek, R., Fonstein, M., D'souza, M., Pusch, G.D., Maltsev, N., 1999. Use of contiguity on the chromosome to predict functional coupling. In Silico Biol. 1 (2), 93–108.

Pal, D., Eisenberg, D., 2005. Inference of protein function from protein structure. Structure 13 (1), 121–130.

Pandey, G., Kumar, V., Steinbach, M., 2006. Computational approaches for protein function prediction: a survey: Technical Report, TR06-028. University of Minnesota, Minneapolis, MN (accessed 14.03.16)http://cs-dev.umn.edu/sites/cs.umn.edu/files/tech_reports/06-028.pdf.

Pandey, G., Myers, C.L., Kumar, V., 2009. Incorporating functional inter-relationships into protein function prediction algorithms. BMC Bioinf. 10, 142.

Pavlidis, P., Weston, J., Cai, J., Grundy, W.N., 2002. Learning gene functional classifications from multiple data types. J. Comput. Biol. 9 (2), 401–411.

Pellegrini, M., Marcotte, E.M., Thompson, M.J., Eisenberg, D., Yeates, T.O., 1999. Assigning protein functions by comparative genome analysis: protein phylogenetic profiles. Proc. Natl. Acad. Sci. U. S. A. 96 (8), 4285–4288.

Peng, W., Wang, J., Cai, J., Chen, L., Li, M., Wu, F.-X., 2014. Improving protein function prediction using domain and protein complexes in PPI networks. BMC Syst. Biol. 8, 35.

Pilpel, Y., Sudarsanam, P., Church, G.M., 2001. Identifying regulatory networks by combinatorial analysis of promoter elements. Nat. Genet. 29 (2), 153–159.

Piovesan, D., Giollo, M., Ferrari, C., Tosatto, S.C.E., 2015. Protein function prediction using guilty by association from interaction networks. Amino Acids 47, 2583–2592.

Przulj, N., Wigle, D.A., Jurisica, I., 2004. Functional topology in a network of protein interactions. Bioinformatics 20, 340–348.

Rangwala, H., Karypis, G., 2005. Profile-based direct kernels for remote homology detection and fold recognition. Bioinformatics 21 (23), 4239–4247.

Raychaudhari, S., Chang, J.T., Sutphin, P.D., Altman, R.B., 2002. Associating genes with gene ontology codes using a maximum entropy analysis of biomedical literature. Genome Res. 12 (1), 203–214.

Renner, A., Aszodi, A., 2000. High-throughput functional annotation of novel gene products using document clustering. In: Proceedings of Pacific Symposium on Biocomputing, pp. 54–68.

Riley, M., 1998. Systems for categorizing functions of gene products. Curr. Opin. Struct. Biol. 8 (3), 388–392.

Rison, S.C., Hodgman, T.C., Thornton, J.M., 2000. Comparison of functional annotation schemes for genomes. Funct. Integr. Genomics 1 (1), 56–69.

Rost, B., Liu, J., Nair, R., Wrzeszczynski, K.O., Ofran, Y., 2003. Automatic prediction of protein function. Cell. Mol. Life Sci. 60 (12), 2637–2650.

Rubinstein, R., Simon, I., 2005. MILANO—custom annotation of microarray results using automatic literature searches. BMC Bioinf. 6 (1), 12.

Ruepp, A., Zollner, A., Maier, D., Albermann, K., Hani, J., Mokrejs, M., Tetko, I., Güldener, U., Mannhaupt, G., Münsterkötter, M., 2004. The FunCat, a functional annotation scheme for systematic classification of proteins from whole genomes. Nucleic Acids Res. 32 (18), 5539–5545.

Rung, J., Schlitt, T., Brazma, A., Freivalds, K., Vilo, J., 2002. Building and analysing genomewide gene disruption networks. Bioinformatics 18 (Suppl. 2), S202–S210.

Salwinski, L., Eisenberg, D., 2003. Computational methods of analysis of protein–protein interactions. Curr. Opin. Struct. Biol. 13 (3), 377–382.

Salwinski, L., Miller, C.S., Smith, A.J., Pettit, F.K., Bowie, J.U., Eisenberg, D., 2004. The Database of Interacting Proteins: 2004 update. Nucleic Acids Res. 32 (Database issue), D449–D451.

Samanta, M.P., Liang, S., 2003. Predicting protein functions from redundancies in large-scale protein interaction networks. Proc. Natl. Acad. Sci. U. S. A. 100 (22), 12579–12583.

Sandhan, T., Yoo, Y., Choi, J.Y., Kim, S., 2015. Graph pyramids for protein function prediction. BMC Med. Genet. 8 (Suppl 2), S12.

Schlitt, T., Palin, K., Rung, J., Dietmann, S., Lappe, M., Ukkonen, E., Brazma, A., 2003. From gene networks to gene function. Genome Res. 13 (12), 2568–2576.

Schomburg, I., Chang, A., Ebeling, C., Gremse, M., Heldt, C., Huhn, G., Schomburg, D., 2004. BRENDA and the enzyme database: updates and major new developments. Nucleic Acids Res. 32 (Database issue), D431–D433.

Schwikowski, B., Uetz, P., Fields, S., 2000. A network of protein–protein interactions in yeast. Nat. Biotechnol. 18 (12), 1257–1261.

Sharan, R., Ideker, T., Kelley, B., Shamir, R., Karp, R.M., 2005. Identification of protein complexes by comparative analysis of yeast and bacterial protein interaction data. J. Comput. Biol. 12, 835–846.

Sharan, R., Ulitsky, I., Shamir, R., 2007. Network-based prediction of protein function. Mol. Syst. Biol. 3, 88.
Simonis, N., van Helden, J., Cohen, G.N., Wodak, S.J., 2004. Transcriptional regulation of protein complexes in yeast. Genome Biol. 5, R33.
Sjolander, K., 2004. Phylogenomic inference of protein molecular function: advances and challenges. Bioinformatics 20 (2), 170–179.
Stoica, E., Hearst, M., 2006. Predicting gene functions from text using a cross-species approach. In: Proceedings of the 11th Pacific Symposium on Biocomputing (PSB), pp. 88–99.
Strong, M., Mallick, P., Pellegrini, M., Thompson, M.J., Eisenberg, D., 2003. Inference of protein function and protein linkages in *Mycobacterium tuberculosis* based on prokaryotic genome organization: a combined computational approach. Genome Biol. 4 (9), R59.
Swift, S., Tucker, A., Vinciotti, V., Martin, N., Orengo, C., Liu, X., Kellam, P., 2004. Consensus clustering and functional interpretation of gene-expression data. Genome Biol. 5 (11), R94.
Tamames, J., Ouzounis, C., Casari, G., Sander, C., Valencia, A., 1998. EUCLID: automatic classification of proteins in functional classes by their database annotations. Bioinformatics 14 (6), 542–543.
Thornton, J.M., Orengo, C.A., Todd, A.E., Pearl, F.M.G., 1999. Protein folds and functions and evolution. J. Mol. Biol. 293 (2), 333–342.
Tornow, S., Mewes, H.W., 2003. Functional modules by relating protein interaction networks and gene expression. Nucleic Acids Res. 31, 6283–6289.
Trivodaliev, K., Bogojeska, A., Kocarev, L., 2014. Exploring function prediction in protein interaction networks via clustering methods. PLoS ONE 9 (6), e99755.
Troyanskaya, O.G., Dolinski, K., Owen, A.B., Altman, R.B., Botstein, D., 2003. A Bayesian framework for combining heterogeneous data sources for gene function prediction (in *Saccharomyces cerevisiae*). Proc. Natl. Acad. Sci. U. S. A. 100 (14), 8348–8353.
Tsuda, K., Shin, H., Scholkopf, B., 2005. Fast protein classification with multiple networks. Bioinformatics 21 (Suppl. 1), ii59–ii65.
Vazquez, A., Flammini, A., Maritan, A., Vespignani, A., 2003. Global protein function prediction from protein–protein interaction networks. Nat. Biotechnol. 21 (6), 697–700.
Vert, J.-P., 2002. A tree kernel to analyze phylogenetic profiles. Bioinformatics 18 (Suppl. 1), S276–S284.
Wachi, S., Yoneda, K., Wu, R., 2005. Interactome-transcriptome analysis reveals the high centrality of genes differentially expressed in lung cancer tissues. Bioinformatics 21, 4205–4208.
Whisstock, J.C., Lesk, A.M., 2003. Prediction of protein function from protein sequence and structure. Q. Rev. Biophys. 36 (3), 307–340.
Wu, C.H., Yeh, L.-S.L., Huang, H., Arminski, L., Castro-Alvear, J., Chen, Y., Hu, Z.-Z., Ledley, R.S., Suzek, P.K.B.E., Vinayaka, C.R., Zhang, J., Barker, W.C., 2003. The protein information resource. Nucleic Acids Res. 31, 345–347.
Wu, Q., Ye, Y., Ng, M.K., Shen-Shyang Ho, S.-S., Shi, R., 2014. Collective prediction of protein functions from protein-protein interaction networks. BMC Bioinf. 15 (Suppl. 2), S9.
Xie, H., Wasserman, A., Levine, Z., Novik, A., Grebinskiy, V., Shoshan, A., Mintz, L., 2002. Large-scale protein annotation through Gene Ontology. Genome Res. 12 (5), 785–794.
Yanai, I., Derti, A., Delisi, C., 2001. Genes linked by fusion events are generally of the same functional category: a systematic analysis of 30 microbial genomes. Proc. Natl. Acad. Sci. U. S. A. 98 (14), 7940–7945.
Yang, Z., 1997. PAML: a program package for phylogenetic analysis by maximum likelihood. Comput. Appl. Biosci. 13 (5), 555–556.

Yu, G., Zhu, H., Domeniconi, C., Guo, M., 2015. Integrating multiple networks for protein function prediction. BMC Syst. Biol. 9 (Suppl. 1), S3.

Zhang, W., Morris, Q.D., Chang, R., Shai, O., Bakowski, M.A., Mitsakakis, N., Mohammad, N., Robinson, M.D., Zirngibl, R., Somogyi, E., Laurin, N., Eftekharpour, E., Sat, E., Grigull, J., Pan, Q., Peng, W.-T., Krogan, N., Greenblatt, J., Fehlings, M., Van Der Kooy, D., Aubin, J., Bruneau, B.G., Rossant, J., Blencowe, B.J., Frey, B.J., Hughes, T.R., 2004. The functional landscape of mouse gene expression. J. Biol. 3 (5), 21.

Zien, A., Kuffner, R., Zimmer, R., Lengauer, T., 2000. Analysis of gene expression data with pathway scores. In: Proceedings of International Conference on Intelligent Systems for Molecular Biology, 8, pp. 407–417.

CHAPTER 2

Reliability of Protein Interactions

Protein interactions play important roles in various biological processes in which the physiological interactions of many proteins are involved in the construction of biological pathways, such as metabolic and signal transduction pathways. In the past few years, high-throughput technologies have identified a large number of protein–protein interactions (PPIs) for many species and mapped protein interactions to their biological processes (Walhout et al., 2000; Rain et al., 2001). The large amount of protein interaction data is a rich information resource from which a lot of biological knowledge and facts can be discovered or inferred, such as the identification of biological complexes and the prediction of protein functions from protein interactions.

However, the protein interaction data generated from various high-throughput technologies usually suffers from high error rates, and as a consequence the biological knowledge discovered from the data is distorted or incorrect. It therefore becomes vital to assess the quality of protein interaction data and extract reliable protein interactions from the high-throughput data for downstream research.

This chapter presents the background of this research area by introducing some representative approaches for assessing the reliability of PPI data and emphasizes a new semantic reliability (SR) approach. This new approach semantically assesses the reliability of each protein interaction and identifies potential false–positive protein interactions in a data set. The distinct feature of this approach is that, when assessing the interaction reliability for each pair of target interacting proteins, it takes into account the semantic influence between proteins that interact with the target proteins and the semantic influence between the target proteins themselves.

2.1 BACKGROUND

The analyses of high-throughput protein interaction data indicate that existing identified protein interactions usually contain many false–positive interactions, that is, the interactions that never take place in cells. Only 30–50%

New Approaches of Protein Function Prediction from Protein Interaction Networks
http://dx.doi.org/10.1016/B978-0-12-809814-1.00002-9
© 2017 Elsevier Ltd.
All rights reserved.

of the identified interactions are biologically relevant, with only a few overlaps among protein interaction data sets from different resources (Deane et al., 2002). As a consequence, the discovered knowledge or inferred facts from protein interactions, such as the interaction clusters, protein complexes, and predicted protein functions, may be distorted or incorrect.

Various approaches have been proposed to assess the reliability of protein interaction data. Von Mering et al. (2002) and Bader and Hogue (2002) analyzed protein interactions from different data sets and resources. Both concluded that interactions identified by multiple experiments and resources were reliable interactions. However, because of the low overlap or limited coverage of different protein interaction data sets, this method may identify few reliable interactions. Deane et al. (2002) proposed the paralogous verification method (PVM) that exploits the paralogs of two interacting proteins and the corresponding paralogous interactions to verify PPIs. This method is restricted within the cases where two proteins involved in an interaction have homologues. Even for this kind of interacting proteins, only half of the interactions were identified with a high reliability confidence under the homology criterion. Deng et al. (2003), Von Mering et al. (2002) and Sprinzak et al. (2003) proposed methods that used the cellular localization and cellular role properties to assess the reliability of PPIs. These methods exploit the proteins' internal structural information for assessing the reliability of protein interactions without considering the external interactions between proteins.

Since proteins perform their functions in various biological processes via their interactions (Chua and Wong, 2009), other approaches have been developed that exploit information conveyed by protein interactions to assess the reliability of interactions. One of the most representative methods is the 'interaction generality' (IG1) method (Saito et al., 2002). The basic idea of the IG1 method is based on an assumption that an interacting pair of proteins (i.e., an interaction) that appears to have many other interacting partners which have no further interactions is likely to be a false–positive interaction. The IG1 method does not consider the topological properties of the interaction network beyond the target protein pair or interaction. To overcome this limitation, Saito et al. (2003) proposed a new interaction generality (IG2) method which improved the IG1 method to assess the reliability of protein interactions. IG2 exploits network topological classifications to determine how well an interaction fits into the expected topology of a PPI network and uses the principal–component analysis when assessing the interaction reliability. Using the ideas behind the IG1 and IG2 methods, Chen et al. (2005, 2006) proposed a method called interaction reliability by alternative path (IRAP) which used only network topological metrics to rank protein interactions according to their computed reliability scores and then identify the reliable interactions. Other approaches tried to integrate existing methods or models to assess the reliability of interactions, such as the probabilistic decision tree (Zhang et al., 2004), logistic regression model (Bader et al., 2004), Naïve Bayes (Jansen et al., 2003), Bayesian network models (Goldberg and Roth, 2003) and Markov random walking clustering (Kamburov et al., 2012). All of these approaches exploit only the topological structure information of a protein interaction network to assess the interaction reliability. However, since the reliability of the interactions used to create an interaction network is still uncertain and yet to be assessed, using only the network structural information from the original interaction data sets to assess the interaction reliability does have inevitable limitations.

Another type of algorithms that make use of the information conveyed by protein interactions to assess the interaction reliability is to incorporate the topological information of interaction networks, as well as the latent semantic information within networks, into the reliability assessment, trying to make the assessment more reasonable. Lin et al. (2009) proposed a Bayesian network-based method that integrated multiple microarray data sets and gene ontology (GO) (GO Consortium, 2016) information to assess the reliability of protein interactions. Wu et al. (2010) also took an integrative way to assess interaction reliability by integrating more types of information, including topological information of protein interaction networks, gene expression files, protein domains and their interactions and protein sequences and their similarity. The limitation of this type of algorithm is that the reliability assessment quality largely depends on the quality of diverse data or information sources, as well as the models or methods used for integration.

A proper way of assessing interaction reliability is to keep a balance between the uncertainty of interaction network and the noises from various data sources in the assessment. Compared with the methods that use interaction topological information to assess reliability, the methods that make use of semantic information between proteins are more reasonable and acceptable, as the semantic information reveals the biological relationships between proteins and can reduce the impact of network uncertainty on the assessment. The method proposed in (Mahdavi and Lin, 2007) primarily tried to assess the interaction reliability semantically by defining a set of rules and selecting eight top ranked GO terms that are extracted from some experimentally confirmed protein interactions in two model organisms and applying the rules and extracted GO terms to assess the interaction reliability. In this method, GO terms are actually used as labels to assess the reliability by checking whether the interacting proteins are labelled by the same GO term(s).

In the flowing sections, we introduce a new SR method that assesses the reliability of protein interactions from the common cellular roles they share and the functional similarity they have and, in turn, extract reliable interactions from original interaction data sets for reliable knowledge discovery. This method properly integrates the protein interaction network topological information with the semantic information of proteins in the reliability assessment. Actually, it uses the protein interaction network as a context to assess the SR of protein interactions. For an interaction between two target proteins, SR method assesses the reliability of the interaction by a reliability score that combines two reliability measures: one is the SR measure between proteins that interact with the target proteins; another is the SR measure between the target proteins themselves. Reliabilities of all interactions in an interaction data set are assessed and sorted by their reliability scores. The interactions with low reliability scores are identified as unreliable and removed from the protein interaction data set, and the remaining interactions form a new reliable interaction data set.

2.2 SEMANTIC INTERACTION RELIABILITY 1 (*R*1)

For an interaction between two proteins, also known as the target interaction and target proteins, respectively, this reliability measure is based on the semantic relationship between the neighbour proteins of two target proteins of an interaction. Therefore this measure is regarded as an *external* reliability measure of an interaction. Neighbour proteins of a target protein are those that directly interact with the target protein.

The semantic relationship between two proteins is determined by the functions of proteins involved in the interaction. SR method adopts the Function Catalogue (FunCat) (Ruepp et al., 2004) scheme for protein function annotation. With FunCat a protein function is expressed in up to six digital layers, and the number of each layer represents a specific function category. For example, the FunCat notation '01.03.01.01' represents the function 'purine nucleotide/nucleoside/nucleobase catabolism', while the notation '01.03.01.01.03' represents the function 'xanthine catabolism'. For a FunCat function notation, the lower-level number represents a subfunction of the higher-level function.

Denote the set of all protein interactions in a data set as R, the set of all proteins involved in the interactions of R as S, and an interaction as a pair of proteins (a, b), where a and b are known as the target proteins of the interaction, that is, $a \in S$, $b \in S$, and $(a, b) \in R$. For a protein $p \in S$, its neighbour proteins (or simply say neighbours) are represented as a set of proteins

$$N_p = \{q \in S | q \text{ has a direct interaction with } p\}$$

For an interaction (a, b), the sets of neighbour protein functions of the proteins a and b are defined respectively as follows:

$$F_a = \{\text{functions of } p | p \in (N_a - \{b\})\}, F_b = \{\text{functions of } p | p \in (N_b - \{a\})\}$$

Then the semantic interaction reliability 1 ($r1$) of an interaction (a, b) is defined as

$$r1(a, b) = \frac{|(F_a \cup F_b) - (F_a \cap F_b)|}{|F_a \cap F_b|} \tag{2.1}$$

where $|\bullet|$ stands for the size of a set, that is, the number of set elements. The numerator in Eq. (2.1) is the number of unmatched functions among the neighbour functions of a and b, while the denominator is the number of matched functions among the neighbour functions of a and b. Therefore this reliability measurement of an interaction (a, b) is actually determined by the ratio of unmatched and matched functions among the neighbour functions of the target proteins a and b. The value of $r1(a, b)$ is within the range $[0, \infty]$. The lower the $r1$ value, the higher the reliability of the interaction. Fig. 2.1 illustrates the calculation of $r1$ for an interaction (a, b).

The reliability measure Eq. (2.1) is based on the neighbour functional information of two target proteins. It is normalized across the whole interaction data set R as a normalized semantic interaction reliability 1, $R1(a, b)$, which is defined as:

$$R1(a, b) = \begin{cases} \dfrac{r1(a, b) - r1_{\min}}{r1_{\max} - r1_{\min}} & \text{if } r1(a, b) \neq \infty \\ 1 & \text{if } r1(a, b) = \infty \end{cases} \tag{2.2}$$

where

$$r1_{\max} = \max\{r1(a, b) | r1(a, b) \neq \infty, (a, b) \in R\}$$
$$r1_{\min} = \min\{r1(a, b) | (a, b) \in R\}$$

The value of $R1(a, b)$ is within the range [0,1].

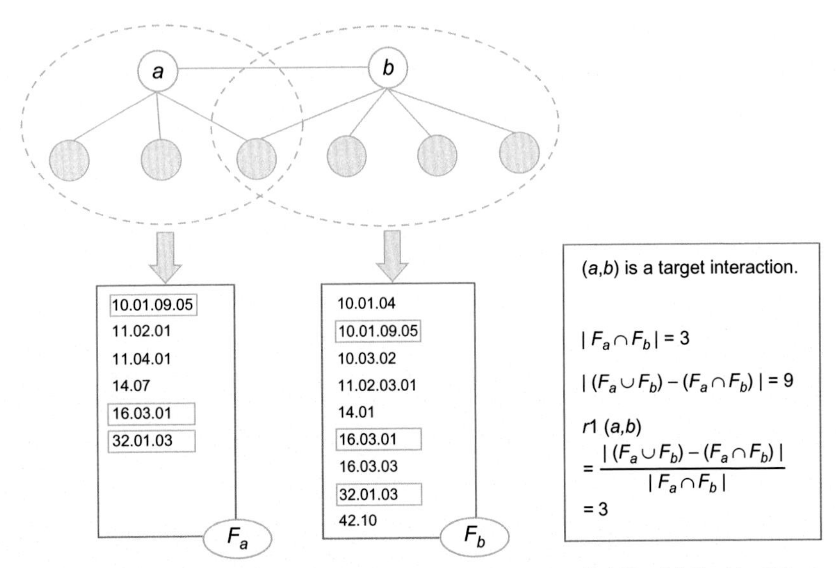

Fig. 2.1 Illustration of calculating the semantic interaction reliability 1 ($r1$). $r1(a, b)$ is the ratio of unmatched and matched functions among the neighbour functions of the target proteins a and b.

2.3 SEMANTIC INTERACTION RELIABILITY 2 (*R*2)

Different from $R1$, the semantic interaction reliability 2 ($R2$) measures the reliability of an interaction by directly measuring the functional similarity of two target proteins. It is based on the assumption that if two target proteins are functionally similar, their interaction is more reliable because the target proteins are more likely to be in the same functional complex and cooperate with each other in biological processes. Therefore $R2$ is regarded as the *internal* reliability measure of an interaction.

In $R2$ a similarity between two protein functions is defined with the FunCat annotation scheme. Based on this function similarity definition, $R2$ between two target proteins is defined. In fact, for any two protein functions f and g that are annotated by the FunCat scheme, their similarity $S_{f,g}$ is defined as

$$S_{f,g} = \frac{\text{number of consecutive matched functional layers of } f \text{ and } g}{\text{maximum number of functional layers of } f \text{ and } g}$$

The function similarity $S_{f,g}$ defines the weight of matched functional layers of two protein functions. For example, suppose function $f = 10.01.05.03.01$ and $g = 10.01.05$, then the number of consecutive matched functional layers is 3, that is, 10.01.05, and the maximum number of functional layers is 5, therefore $S_{f,g}=3/5=0.6$.

For a protein $p \in S$, denote the set of functions protein p possesses as $F(p)$. For a protein interaction (a, b) where a and b are two target proteins, the semantic interaction reliability 2 ($r2$) of the interaction is defined as

$$r2(a,b)=\max\left(\sum_{f\in F(a)}\left(\max_{g\in F(b)} S_{f,g}\right), \sum_{g\in F(b)}\left(\max_{f\in F(a)} S_{g,f}\right)\right) \tag{2.3}$$

The first element $\sum_{f\in F(a)}\left(\max_{g\in F(b)} S_{f,g}\right)$ in Eq. (2.3) measures the functional similarity from the point of view of target protein a, while the second element $\sum_{g\in F(b)}\left(\max_{f\in F(a)} S_{g,f}\right)$ measures the functional similarity from the point of view of target protein b. The value of $r2(a, b)$ is within the range $[0, \infty)$, and the higher the $r2$ value, the higher the reliability of the interaction.

The reliability measurement Eq. (2.3) is based on the local functional information of two target proteins. Similar to $r1$, it is normalized across the whole interaction data set R as a normalized semantic interaction reliability 2, $R2(a, b)$, which is defined as:

$$R2(a,b)=\frac{r2(a,b)-r2_{\min}}{r2_{\max}-r2_{\min}} \tag{2.4}$$

where

$$r2_{\max}=\max\{r2(a,b)|(a,b)\in R\}$$
$$r2_{\min}=\min\{r2(a,b)|(a,b)\in R\}$$

The value of $R2(a, b)$ is within the range [0,1].

The calculation of $R2$ relies on the condition that the functions of two target proteins a and b are known. If one of the target proteins (a or b) or both target proteins (a and b) are unannotated (i.e., their functions are unknown), the majority voting method (Schwikowski et al., 2000) is used to assign functions to the target protein(s); that is, functions that occur most frequently in the neighbours of the target protein are assigned to the target protein as its functions.

2.4 SEMANTIC RELIABILITY (SR) ASSESSMENT

Based on the semantic interaction reliability measures $R1$ and $R2$ which respectively measure the reliability of a protein interaction externally and internally, a new semantic reliability measure (*SR*) is defined that combines these two reliability measures to assess the overall reliability of an interaction and, in turn, to remove unreliable interactions from an interaction data set.

For an interaction (a, b), its semantic reliability $SR(a, b)$ is defined as the logarithmic value of the ratio between $R2$ and $R1$. This SR is used to measure the overall reliability of an interaction. Since the values of $R1$ and $R2$ vary in different cases, the calculation of the $SR(a, b)$ varies accordingly. Table 2.1 provides the details for calculating the $SR(a, b)$ values for different cases, where the min values are across the whole interaction data set and the intersections of $R1$ and $R2$ value columns provide the formulas for calculating $SR(a, b)$ values accordingly. The experimental observation (Hou and Saini, 2013) shows that the values of $R1$ and $R2$ for an interaction are independent. Therefore the value of *SR* for an interaction is not overestimated.

By using Table 2.1 each interaction in an interaction data set is assigned a *SR* value. Since proteins implement their functions by locally interacting with their neighbour proteins, the final reliability of an interaction is therefore assessed in the context of local neighbours. To this end an average SR of protein a within the domain of local neighbours is defined as

$$AR_a = \frac{\sum_{p \in N_a} SR(a, p)}{|N_a|}$$

where $|N_a|$ stands for the number of neighbour proteins of protein a. Accordingly, whether an interaction (a, b) is reliable or not is determined by the following algorithm:

If $SR(a, b) \geq AR_a$ or $SR(a, b) \geq AR_b$
then the interaction (a, b) is reliable;
else
the interaction (a, b) is unreliable.

Table 2.1 Calculation of SR of an interaction for different *R*1 and *R*2 values

		***R*1**	
		0	**(0, 1]**
*R*2	0	$\ln\left(\frac{\min_{R2 \neq 0} R2}{\min_{R1 \neq 0} R1}\right)$	$\ln\left(\frac{\min_{R2 \neq 0} R2}{R1}\right)$
	(0,1]	$\ln\left(\frac{R2}{\min_{R1 \neq 0} R1}\right)$	$\ln\left(\frac{R2}{R1}\right)$

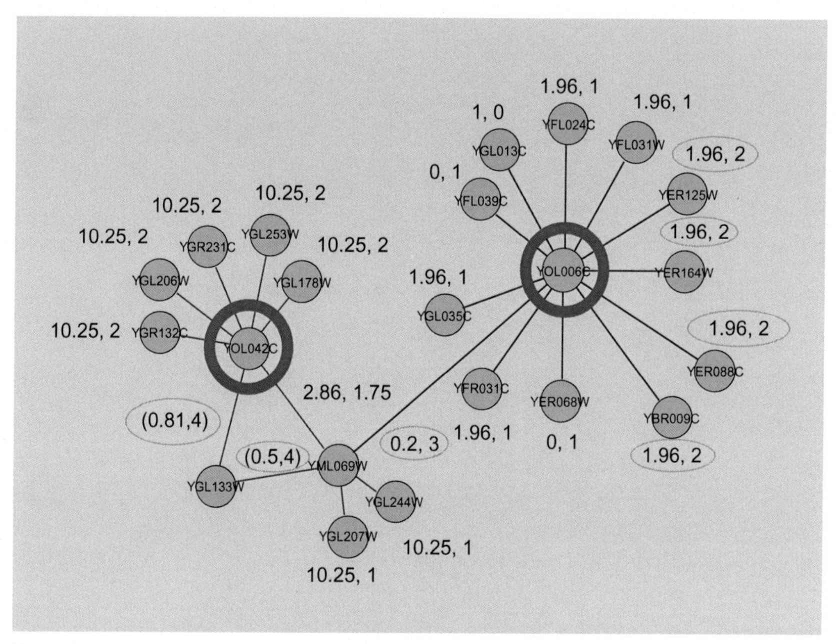

Fig. 2.2 A part of an original protein interaction network.

If an interaction in a data set is identified as unreliable by the preceding algorithm, it is removed/filtered from the data set. Figs. 2.2 and 2.3 (Hou and Saini, 2013) respectively show a part of an original protein interaction data set and the filtered data set generated from the preceding algorithm.

In Fig. 2.2 the network has two core proteins, YOL006c and YDL042c (highlighted in thick circles). The $R1$ and $R2$ values of each interaction are marked in the format ($R1$, $R2$). The interactions with $R1$ and $R2$ values in the thin oval-circles are reliable according to the algorithm. After the unreliable (false–positive) interactions in the original interaction network have been removed, a new reliable protein interaction network is generated, which is shown in Fig. 2.3.

2.5 DISCUSSIONS

The SR measure presented in the preceding section exploits the semantic relationship between protein functions to assess the reliability of protein interactions. For a protein interaction, this measure combines the external ($R1$) and internal ($R2$) reliabilities of the interaction to assess the overall SR of the interaction. Accordingly, the unreliable interactions could be

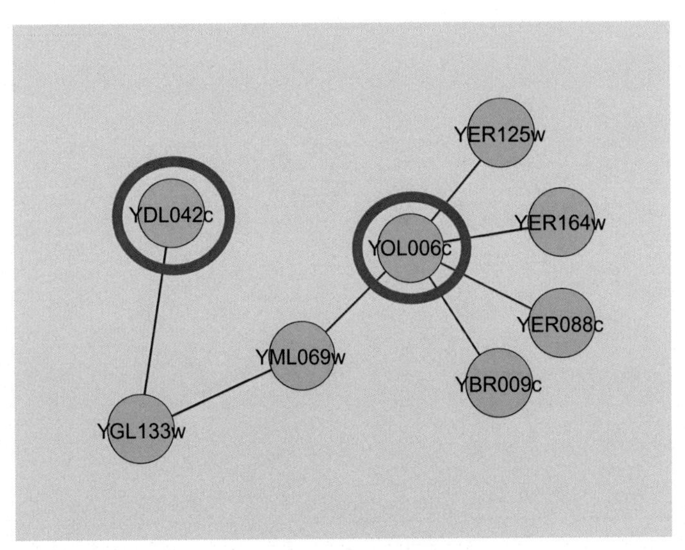

Fig. 2.3 The filtered protein interaction network of Fig. 2.2.

removed or filtered from the original protein interaction network to generate a more reliable network.

In the reliability measures *R*1 and *R*2, that is, Eqs. (2.2), (2.4), different approaches are taken to exploit functional information to measure the reliability. In *R*1 the measure requires functions to be exactly matched when checking whether the functions are shared by the neighbour proteins, while in *R*2 the measure does not require the functions of two target proteins to be exactly matched, but does require the functions to be matched at certain functional levels. This is because in *R*1, before the interactions are assessed, some proteins may not be the real neighbours of the target protein. Therefore a stricter criterion is applied; that is, functions must be exactly matched when checking whether functions are shared by neighbours so that the impact of noisy neighbour proteins on the reliability measurement can be reduced. While for two target proteins in the *R*2 measure, the approach is to examine at which functional levels the functions of these two proteins are exactly matched, even if the functions are not exactly the same. This approach makes the *R*2 measure more objective and reasonable, because the interacting proteins are more likely to be in a functional category even though their functions are not exactly the same (Chua and Wong, 2009). Combining the *R*1 and *R*2 reliability measures into the semantic reliability measure *SR* more objectively reflects the real situations of original protein interaction data sets.

In addition to these distinct features, *SR* measure, as well as the related measures such as *R*1, *R*2 and function similarity, also provides a framework based on which some new approaches or methods could be proposed to further improve the reliability assessment or other performance such as protein function prediction. For instance, in the *R*2 reliability measure, if functions of the target proteins are unknown, the majority voting method (Schwikowski et al., 2000) is adopted to assign functions to the target proteins because of its simplicity and low cost of computation. It is obvious that other available function prediction methods could also be adapted for the same purpose, which might more effectively assess the interaction reliability.

Furthermore, the issue of adapting an existing function prediction method to assign functions to the unannotated target proteins in *R*2 measure is that the assigned/predicted functions are based on the un-assessed interactions. Therefore assigned functions of the target proteins might not be 100% reliable. On the other hand, if protein interactions are reliable, the assigned/predicted functions are most likely to be reliable. In other words, the reliabilities of the interactions and the assigned functions depend on each other. This fact indicates a possible research direction for improving the interaction reliability as well as function prediction of unannotated proteins: incorporating the interaction reliability assessment and function prediction into an iterative process which enables the reliability assessment and function prediction to enhance/endorse each other iteratively to finally achieve the better results.

The function notation and similarity in SR measure both are based on the FunCat (Ruepp et al., 2004) annotation scheme. It is obvious that the semantic similarity between two functions in the *R*2 reliability measure could also be calculated by using other available methods such as the one in Zhu et al. (2010). It is also possible to annotate function and define function similarity using other schemes such as the GO (GO Consortium, 2016), and incorporate them into *R*1, *R*2 and *SR* calculations for reliability assessment. It is worth investigating the impact of other function similarity definitions on the effectiveness of reliability assessment methods.

The reliability assessment methods introduced in this chapter mainly focus on the existing reported protein interaction networks. Actually, reliability assessment could also be an integrated part of in silico protein interaction prediction, as interaction prediction is based on the judgement of interaction reliability from other information resources. Integrating interaction reliability assessment with interaction prediction is beyond the scope of this book. Interested readers can find a substantial amount of literature about protein interaction prediction.

Although various algorithms have been proposed to eliminate unreliable interactions in order to get a better quality protein interaction data set for downstream research such as protein function prediction, no algorithms can guarantee all interactions in the new filtered data set are one hundred percent reliable. So regarding the protein function prediction from protein interaction networks, another possible research direction is to allow the existence of unreliable interactions in the data sets and make the prediction algorithm capable of reducing or removing the influence of unreliable interactions on the prediction results. For example, algorithms could be developed to iteratively predict function rather than once-off to dynamically select domains for function prediction, or to incorporate semantic information into prediction procedures and so on. These new techniques will be introduced in the succeeding chapters.

2.6 EVALUATION OF RELIABILITY ASSESSMENT

Evaluation of reliability assessment is vital for effective judgement of reliability assessment methods. The commonly used evaluation methods can be generally classified into two categories: direct and indirect evaluations. The direct category can be further divided into three subcategories: biological experimental evaluation, literature evaluation and computational evaluation.

Biological experimental evaluation tries to verify the reproducibility of the interactions that are assessed as reliable by computational methods (Lin et al., 2009; Chua and Wong, 2008). The idea of this type of evaluation is based on a reasonable assumption that a reliable interaction should be reproducible in more than one biological experiment, and the more experiments confirming it, the more confidence the interaction is reliable. This kind of evaluation can be conducted by designing and carrying out new biological experiments, which are usually high in cost or by checking against the confirmed reliable interactions in the existing interaction databases, such as MIPS (Mewes et al., 2002), DIP (Salwinski et al., 2004) and BioGRID (Chatr-Aryamontri et al., 2015). However, the issue with this kind of evaluation is that data produced by different technologies do not overlap significantly, and also the data produced at different labs using the same technologies differ substantially (Chen and Xu, 2003). This fact suggests that these databases might provide complementarities to each other.

Literature evaluation is based on literature mining and an assumption that if two protein names appear in the same article, they have a chance to interact with each other. If the two protein names coexist in more articles, their interaction is more reliable. The existing databases, such as the gene-to-gene co-citation network for 13,712 named human genes generated from literature mining over 10 million MEDLINE records (Jenssen et al., 2001), make it possible to evaluate the computational assessed reliable interactions using a wide range of available literature.

Computational evaluation relies on other reliability assessment results provided by other computational methods or on other computational measures to evaluate and validate the predicted reliabilities. If a predicted reliable interaction is confirmed by more prediction methods, it is more reliable. A commonly used computational measure is the functional similarity scores between proteins based on the GO terms (Wang et al., 2007). This computational measure is often used for computational validation of protein interactions (Wu et al., 2010; Chen et al., 2006; Pei and Zhang, 2005).

Indirect evaluation is to use other evidence to evaluate the reliability assessment or to compare the performance results (before and after the predicted unreliable interactions being removed) of other algorithms that utilize the protein interaction data sets to obtain results, such as predicted functions. The evidence could be some semantic information from other sources such as GO. As discussed by Sprinzak et al. (2003), for true interactions, the interacting proteins should be localized in the same cellular compartment or participate in the same cellular process. Therefore a predicted reliable interaction can be evaluated by considering the shortest distance of the two target proteins in GO cellular component, molecular function, and biological process to see if they meet these true interaction conditions (Lin et al., 2009). Another kind of evidence could be the performance properties of the reliability assessment algorithm in terms of commonly used measures, such as the Area Under the ROC Curve (AUC) and the Receiver Operating Characteristic (ROC) curve, to indirectly confirm that the predicted reliable interactions are trustable in most cases by comparing them with other computational reliability prediction methods (Lin et al., 2009; Wu et al., 2010; Kamburov et al., 2012).

Hou and Saini (2013) evaluated the predicted reliable interactions indirectly by examining the performance of representative function prediction algorithms on the unfiltered and filtered interaction data sets generated by the semantic reliability assessment method SR. In addition, the effectiveness of the SR method was evaluated by comparing it with other representative reliability assessment methods. Particularly, the SR and other comparing methods were used to generate filtered data sets from the same original unfiltered data sets. Then some representative function prediction methods were applied on these filtered data sets to generate different function prediction results, and the effectiveness of the reliability assessment methods was evaluated by the reliability of the predicted functions. The idea behind this evaluation approach was that the existing protein function prediction methods that exploit protein interaction data to predict protein functions were based on an assumption that the available protein interactions were reliable. Therefore, if a protein function prediction method can achieve the higher prediction precision, recall and *F*-value from a filtered protein interaction data set, it implies that this filtered data set contains more reliable interactions. Accordingly, the reliable assessment method that generates this filtered data set is more effective, and the identified reliable interactions are more trustable in most cases.

REFERENCES

Bader, G.D., Hogue, C.W., 2002. Analysing yeast protein–protein interaction data obtained from different sources. Nat. Biotechnol. 20 (10), 991–997.

Bader, J.S., Chaudhuri, A., Rothberg, J.M., Chant, J., 2004. Gaining confidence in high-throughput protein interaction networks. Nat. Biotechnol. 22 (1), 78–85.

Chatr-Aryamontri, A., Breitkreutz, B.J., Oughtred, R., Boucher, L., Heinicke, S., Chen, D., Stark, C., Breitkreutz, A., Kolas, N., O'Donnell, L., Reguly, T., Nixon, J., Ramage, L., Winter, A., Sellam, A., Chang, C., Hirschman, J., Theesfeld, C., Rust, J., Livstone, M.S., Dolinski, K., Tyers, M., 2015. The BioGRID interaction database: 2015 update. Nucleic Acids Res. 43 (Database issue), D470–D478.

Chen, Y., Xu, D., 2003. Computational analyses of high-throughput protein–protein interaction data. Curr. Protein Pept. Sci. 4, 159–181.

Chen, J., Hsu, W., Lee, M.L., Ng, S.-K., 2005. Discovering reliable protein interactions from high-throughput experimental data using network topology. Artif. Intell. Med. 35, 37–47.

Chen, J., Hsu, W., Lee, M.L., Ng, S.-K., 2006. Increasing confidence of protein interactomes using network topological metrics. Bioinformatics 22 (16), 1998–2004.

Chua, H.N., Wong, L., 2008. Increasing the reliability of protein interactomes. Drug Discov. Today 13 (15/16), 652–658.

Chua, H.N., Wong, L., 2009. Predicting protein functions from protein interaction networks. Biological Data Mining in Protein Interaction Networks. IGI Global, Hershey, PA, pp. 203–222.

Deane, C.M., Salwinski, L., Xenarios, I., Eisenberg, D., 2002. Protein interactions: two methods for assessment of the reliability of high-throughput observations. Mol. Cell. Proteomics 1 (5), 349–356.

Deng, M., Zhang, K., Mehta, S., Chen, T., Sun, F., 2003. Prediction of protein function using protein–protein interaction data. J. Comput. Biol. 10 (6), 947–960.

GO Consortium, 2016. Ontology Structure. http://geneontology.org/page/ontology-structure (accessed 11.03.16).

Goldberg, D.S., Roth, F.P., 2003. Assessing experimentally derived interactions in a small world. Proc. Natl. Acad. Sci. U. S. A. 100 (8), 4372–4376.

Hou, J., Saini, A., 2013. Semantically assessing the reliability of protein interactions. Math. Biosci. 245 (2), 226–234.

Jansen, R., Yu, H., Greenbaum, D., Kluger, Y., Krogan, N.J., Chung, S., Emili, A., Snyder, M., Greenblatt, J.F., Gerstein, M., 2003. A Bayesian networks approach for predicting protein–protein interactions from genomic data. Science 302 (5644), 449–453.

Jenssen, T.K., Laegreid, A., Komorowski, J., Hovig, E., 2001. A literature network of human genes for high-throughput analysis of gene expression. Nat. Genet. 28 (1), 21–28.

Kamburov, A., Grossmann, A., Herwig, R., Stelzl, U., 2012. Cluster-based assessment of protein–protein interaction confidence. BMC Bioinform. 13, 262.

Lin, X., Liu, M., Chen, X.-W., 2009. Assessing reliability of protein–protein interactions by integrative analysis of data in model organisms. BMC Bioinform. 10 (Suppl. 4), S5.

Mahdavi, M.A., Lin, Y.-H., 2007. False positive reduction in protein–protein interaction predictions using gene ontology annotations. BMC Bioinform. 8, 262.

Mewes, H.W., Frishman, D., Guldener, U., Mannhaupt, G., Mayer, K., Mokrejs, M., Morgenstern, B., Munsterkotter, M., Rudd, S., Weil, B., 2002. MIPS: a database for genomes and protein sequences. Nucleic Acids Res. 30 (1), 31–34.

Pei, P., Zhang, A., 2005. A topological measurement for weighted protein interaction network. Proc. IEEE Comput. Syst. Bioinform. Conf. 2005, 268–278.

Rain, J.C., Selig, L., Reuse, H.D., Battaglia, V., Reverdy, C., Simon, S., Lenzen, G., Petel, F., Wojcik, J., Schachter, V., Chemama, Y., Labigne, A., Legrain, P., 2001. The protein–protein interaction map of *Helicobacter pylori*. Nature 409 (6817), 211–215.

Ruepp, A., Zollner, A., Maier, D., Albermann, K., Hani, J., Mokrejs, M., Tetko, I., Güldener, U., Mannhaupt, G., Münsterkötter, M., 2004. The FunCat, a functional annotation scheme for systematic classification of proteins from whole genomes. Nucleic Acids Res. 32, 5539–5545.

Saito, R., Suzuki, H., Hayashizaki, Y., 2002. Interaction generality, a measurement to assess the reliability of a protein–protein interaction. Nucleic Acids Res. 30 (5), 1163–1168.

Saito, R., Suzuki, H., Hayashizaki, Y., 2003. Construction of reliable protein–protein interaction networks with a new interaction generality measure. Bioinformatics 19 (6), 756–763.

Salwinski, L., Miller, C.S., Smith, A.J., Pettit, F.K., Bowie, J.U., Eisenberg, D., 2004. The database of interacting proteins: 2004 update. Nucleic Acids Res. 32 (Database issue), D449–D451.

Schwikowski, B., Uetz, P., Fields, S., 2000. A network of interacting proteins in yeast. Nat. Biotechnol. 18, 1257–1261.

Sprinzak, E., Sattath, S., Margalit, H., 2003. How reliable are experimental protein–protein interaction data? J. Mol. Biol. 327 (5), 919–923.

Von Mering, C., Krause, R., Snel, B., Cornell, M., Oliver, S.G., Fields, S., Bork, P., 2002. Comparative assessment of large-scale data sets of protein–protein interactions. Nature 417 (6887), 399–403.

Walhout, A.J.M., Boulton, S.J., Vidal, M., 2000. Yeast two-hybrid systems and protein interaction mapping projects for yeast and worm. Yeast 17 (2), 88–94.

Wang, J., Du, Z., Payattakool, R., Yu, P., Chen, C., 2007. A new method to measure the semantic similarity of GO terms. Bioinformatics 23 (10), 1274–1281.

Wu, M., Li, X., Chua, H.N., Kwoh, C.-K., Ng, S.-K., 2010. Integrating diverse biological and computational sources for reliable protein–protein interactions. BMC Bioinform. 11 (Suppl. 7), S8.

Zhang, L.V., Wong, S.L., King, O.D., Roth, F.P., 2004. Predicting co-complexed protein pairs using genomic and proteomic data integration. BMC Bioinform. 5, 38.

Zhu, W., Hou, J., Chen, Y.-P., 2010. Semantic and layered protein function prediction from PPI networks. J. Theor. Biol. 267 (2), 129–136.

CHAPTER 3

Clustering-Based Protein Function Prediction

The large amount of protein–protein interaction (PPI) data generated by high-throughput technologies for many model species makes it possible to use computational approaches to predict functions of unannotated proteins. Usually PPI data can be modelled as a network where a node represents a protein and an edge between two nodes represents an interaction between two proteins. A protein interaction network provides a context within which functions of unannotated proteins could be predicted from the available functions possessed by other proteins in the protein interaction network. Actually, to effectively predict functions from PPI networks, computational approaches have to address the following issues:

(i) How to validate the reliability of PPI data, how to filter noise information from the existing data sets, or how to reduce the impact of noise information on the prediction results?

(ii) How to reasonably measure the protein similarity or distance from protein interactions? This is the key to more precisely and reasonably discovering the biological patterns and characteristics of proteins.

(iii) How to propose effective and efficient algorithms for function prediction?

(iv) To what extent or level should the multiple functions be predicted for unannotated proteins?

Chapter 2 introduced some solutions for the first issue. This chapter focuses on the issues (ii), (iii), and (iv) by introducing a new sematic similarity between proteins, a new clustering-based approach to predict protein functions at different biological levels, and a new dynamic clustering algorithm for more effectively predicting functions.

3.1 BACKGROUND

Clustering is a commonly used and effective module-based approach for protein function prediction from PPI networks (Brohée and van Helden, 2006).

New Approaches of Protein Function Prediction from Protein Interaction Networks
http://dx.doi.org/10.1016/B978-0-12-809814-1.00003-0
© 2017 Elsevier Ltd.
All rights reserved.

This approach is based on the assumption that proteins in the same cluster are more likely to share some common or similar functions. Therefore, if an unannotated protein is clustered into a cluster with other annotated proteins, it is possible to use the functions of these annotated proteins to predict the functions of the unannotated protein. In addition, a clustering approach is capable of exploiting the direct, as well as the indirect, interaction proteins of the unannotated protein to predict functions, and of more effectively interpreting the protein interaction relationships in the prediction process.

Various clustering methods have been proposed to identify protein modules or complexes or to predict protein functions from protein interaction networks (Enright et al., 2002; Brun et al., 2003; Samanta and Liang, 2003; Brun et al., 2004; King et al., 2004; Pereira-Leal et al., 2004; Przulj et al., 2004; Arnau et al., 2005; Dunn et al., 2005; Maciag et al., 2006; Adamcsek et al., 2006; Brohée and van Helden, 2006; Chen and Yuan, 2006; Luo et al., 2007; Asur et al., 2007; Trivodaliev et al., 2014). Although the existing methods can obtain satisfactory results in some cases, no methods are applicable to general cases. This is because detecting clusters generally is very hard to do and is not yet satisfactorily resolved (Fortunato, 2010). Actually, the use of clustering methods to predict protein function from PPI networks requires answering two questions: one is how to measure the relationship between proteins, especially in terms of similarity or distance; another is what clustering method should be used and how to predict functions from the generated clusters. Regarding the similarity or distance measurement between proteins, the better a measure reasonably and properly reflects the semantic or functional relationship between proteins, the better the clustering and prediction results will be, as long as a proper clustering method is applied. It is obvious that these are the targets we expect to achieve, although there is still a long way to go.

Clustering-based methods predict functions of an unannotated protein from the functions of clusters that are related to the unannotated protein. Functions in a cluster might show different spectrums, some functions might be the same, some might be different but still belong to the same functional category, or some might be different but complementary to each other in a biological process. The variety of functions in a cluster makes it possible to predict functions of an unannotated protein at different functional levels using clustering-based methods. In biological research and applications, predicting protein function at different functional levels is more applicable for some reasons: the interaction information in PPI data is not 100% correct, the real biological and functional relationship between proteins could not be revealed by PPI networks, and the limitations of computational methods make it impossible to ensure that the prediction results exactly match the real functions. So predicting functions at different functional levels helps us understand protein functions from different angles and narrow the investigation area when designing biological experiments. This chapter introduces a new clustering-based approach to predict functions of unannotated proteins at different functional levels or layers. Meanwhile, a new semantic similarity between proteins is introduced, which is the base of this new prediction approach.

On the other hand, clustering proteins for function prediction is a dynamic process with which the possible predicted functions are grouped into the clusters that are related to the unannotated proteins. However, protein interaction data sets usually contain many false positive interactions (Deane et al., 2002), which might result in the irrelevant proteins being improperly clustered into the clusters from which functions are predicted. Although existing methods are developed to assess the reliability of interactions and remove unreliable interactions from a PPI data set, no methods can guarantee that all false positive interactions can be identified and removed. In other words, simply predicting protein functions from the final generated clusters without considering the dynamic features of clustering processes may significantly lower the prediction quality.

Another issue about the quality of clustering-based prediction is the selection of a proper clustering method and the ways of using it for effective function prediction, although a lot of clustering methods have been developed in a large interdisciplinary scientific community. This also refers to how to select feature functions of each cluster for function prediction. For example, if the *k*-means clustering method is used, the value of the clustering parameter *k* has an impact on the quality of clustering and, in turn, on the quality of the function prediction. Similarly, if a hierarchical clustering algorithm is used, the extent to which the clustering is conducted and the way of predicting functions from clusters have a significant impact on the quality of clustering and function prediction. These issues must be addressed and resolved when we use clustering-based approaches to predict functions of unannotated proteins.

To address the preceding issues, along with the new layered function prediction method, this chapter also introduces some new methods of selecting feature functions of a cluster for effective function prediction. Furthermore, a new clustering-based method is introduced to show a new way of progressively using a clustering method to predict functions. This method is innovative in several aspects. In addition to considering the final generated clusters when predicting functions, this method dynamically traces the appearance of the functions across all generated clusters that are relevant to the unannotated proteins when progressively clustering the proteins, and selects those functions as the predicted functions that have a higher stability in this progressive clustering process.

Since protein similarity is the base of protein clustering and, in turn, the base of clustering-based function prediction, a new semantic protein similarity is firstly introduced and then it is followed by the clustering-based function prediction methods in the upcoming sections.

3.2 SEMANTIC PROTEIN SIMILARITY

Different from many existing protein similarity measures in the literature that are based on gene ontology (GO) terms (GO Consortium, 2016), the new semantic protein similarity introduced in this section is based on another commonly used functional annotation scheme FunCat (Ruepp et al., 2004). With the hierarchical FunCat scheme, a function can be expressed up to six biological layers at most. The biological function category at each layer is represented as a two-digit number, and a function is annotated by a series of two-digit numbers (function categories) concatenated by a dot (.). The more layers a function notation has, the more specific the function is. For example, the notation '01.01.06.06.01', which has five layers, denotes the function '*biosynthesis of lysine*', which is more specific than the function '*metabolism of lysine*' denoted by '01.01.06.06', which has four layers.

With the layered structure of the FunCat scheme, the first layer of a function annotation is defined as the first two-digit number from the left, the second layer is the second two-digit number from the left and so on until the sixth layer. Since the semantic information each layer carries is different (i.e., the deeper a layer, the more specific a function), in order to more specifically measure the semantic importance of each function annotation layer, the weight of the l-th layer ($l \in [1,6]$) is defined as

$$w(l) = \frac{\sum_{k=1}^{l} k^2}{\sum_{n=1}^{6} n^2} \tag{3.1}$$

For example, the weight of the first layer is $w(1) = 1/91 = 0.011$, the weight of the third layer is $w(3) = (1+4+9)/91 = 0.1538$, and the weight of the sixth layer is $w(6) = 1$. The parameter k^2 in the preceding formula is defined to give more weight to a deeper layer because a deeper layer carries more specific and semantic information about a function.

Based on the preceding layer weight definition, a semantic similarity between two annotated proteins can be defined from their functional comparisons. The similarity depends on two factors: one is the number of valid functional comparisons; another is the matched function levels of each valid functional comparison. If two functions are not the same at level 1, the functions are then considered irrelevant, and the comparison between these two functions is considered invalid; otherwise the comparison is valid. Suppose there are two proteins, i and j, with m and n functions, respectively, then the number of all possible functional comparisons between these two proteins is $N_c(i, j) = m \times n$. Denote the number of invalid comparisons as $N_z(i,j)$, then the number of valid functional comparisons $VC(i,j)$ between proteins i and j is defined as

$$VC(i, j) = N_c(i, j) - N_z(i, j). \tag{3.2}$$

For any two annotated proteins i and j in a PPI network, their functional semantic similarity $sim(i, j)$ is defined as

$$sim(i, j) = \begin{cases} \sum_{s=1}^{VC(i,j)} w(l_s)/VC(i, j), & VC(i, j) \neq 0 \\ 0, & VC(i, j) = 0 \end{cases} \tag{3.3}$$

where $l_s \in [1, 6]$ is the deepest matched layer of the s-th functional comparison between two proteins i and j, and $w(l_s)$ is the weight of the layer l_s as defined in (3.1). For example, if the s-th functional comparison refers to two functions $f = 01.01.06.06.01$ and $g = 01.01.06.05$ which belong to proteins i and j, respectively, then $l_s = 3$ and $w(l_s) = w(3) = 0.1538$.

Accordingly, the distance between two interacting annotated proteins i and j in a PPI network is defined here by their semantic similarity

$$dis(i, j) = \begin{cases} \dfrac{1}{sim(i, j)} - 1, & sim(i, j) \neq 0 \\ \dfrac{1}{\mu}, & sim(i, j) = 0 \end{cases}, \tag{3.4}$$

where μ is a constant that is small enough, so the distance is therefore big enough, to distinguish the distance between two proteins that are not similar (i.e., $sim(i, j) = 0$) to other normal distances. It can be estimated from the similarity definition (3.3) that the maximum normal distance between two proteins is approximately $10^2-1=99$. Therefore the value of μ can be set as $\mu=10^{-4}$, and the corresponding distance is 10^4, which is great enough to distinguish it from the maximum normal distance (Zhu et al., 2010).

The preceding definition of the new similarity or distance between two annotated proteins in a PPI incorporates all referred function semantics numerically, as well as the significance of each functional layer in the involved functions.

3.3 LAYERED PROTEIN FUNCTION PREDICTION

With the new semantic similarity (3.3) or distance (3.4) of proteins, the function prediction of unannotated proteins can be made now by using clustering methods. Actually, clustering is conducted on all annotated proteins in a PPI network, with their links (i.e., interactions) being converted into similarities or distances according to (3.3) or (3.4). Any available similarity or distance-based clustering method can be used for clustering annotated proteins in this algorithm, such as the clustering methods in de Hoon et al. (2004). So the following function prediction algorithm assumes that all annotated proteins have already been clustered. The functions of a cluster are defined as the functions possessed by the proteins in the cluster. Then function prediction for the unannotated proteins are made based on the feature functions of these clusters and the interaction information between the unannotated proteins and the clusters. This prediction algorithm consists of the following steps:

(i) Cluster function ranking and feature function selection
(ii) Cluster feature function weighting
(iii) Interaction weight calculation between an unannotated protein and its interacting clusters
(iv) Function scoring and prediction

The details of these steps are presented as follows.

Cluster function ranking and feature function selection: Feature functions of a cluster are the most common functions within a cluster. Feature functions of a cluster can be identified at different functional layers, because a function is annotated in, at most, six layers with the FunCat annotation scheme. Ranking functions and then selecting feature functions of a cluster at different functional layers make it possible to predict functions of unannotated proteins at different functional layers, which in turn provides more biological category information and guidelines, rather than just specific functions, which helps to understand the possible biological functions of the unannotated proteins.

Function ranking within a cluster is based on the occurrence frequency of a function in the cluster. A function with a higher frequency is ranked higher. Feature function selection of a cluster depends on two factors: how many functional categories are expected in function prediction and at which functional layer, the predicted functions are expected to be. Denote the number of expected functional categories as *EC* and the number of expected functional layers of predicted functions as *EL*. The details of function ranking and feature function selection of a cluster are as follows.

The function ranking and feature function selection within a cluster start from layer-1 of all functions in the cluster. All layer-1 functions are ranked in a descending order according to their occurrence frequencies in the cluster. Then the highly ranked layer-1 functions are selected, and the number of selected layer-1 functions is *EC*. These selected layer-1 functions represent the most common function categories in the cluster. Then for each selected layer-1 function, all its related layer-2 functions in the cluster are ranked, and the highest ranked one is selected as a feature function of the cluster at layer-2. This process can proceed by ranking all layer-3 functions for each selected layer-2 feature function, selecting the highest ranked function as a feature function of the cluster at layer-3, and so on. The number of layers of the feature functions depends on the value of *EL*, that is, the number of expected functional layers of predicted functions. For example, suppose we set $EC=3$ and $EL=4$, and the most common function categories (layer-1 functions) in a cluster are 01 (*Metabolism*), 11 (*Transcription*), and 12 (*Protein Synthesis*), accordingly. For function 01, if the functions 01.20, 01.20.05, and 01.20.05.07 are the highest ranked functions at layer-2, layer-3, and layer-4 (because of $EL=4$), respectively, then these functions are selected as the feature functions of the cluster at layer-2, layer-3, and layer-4, respectively. For functions with the layer-1 functions 11 and 12, the same ranking and selection operations are conducted. The sample feature functions of a cluster for this example are presented in Fig. 3.1.

The preceding function ranking and feature function selection algorithm provides flexibilities to protein function prediction algorithms for choosing candidate functions and predicting functions of unannotated proteins at different functional layers. Specifically, if a function prediction expects to cover more function categories, a higher value of parameter *EC* can be set in the preceding algorithm. If a function prediction expects to predict functions more specifically (i.e., at a deeper layer), a higher value of parameter *EL* can be set in the preceding algorithm. The combination of the *EC* and *EL* values makes it possible for function prediction algorithms to predict

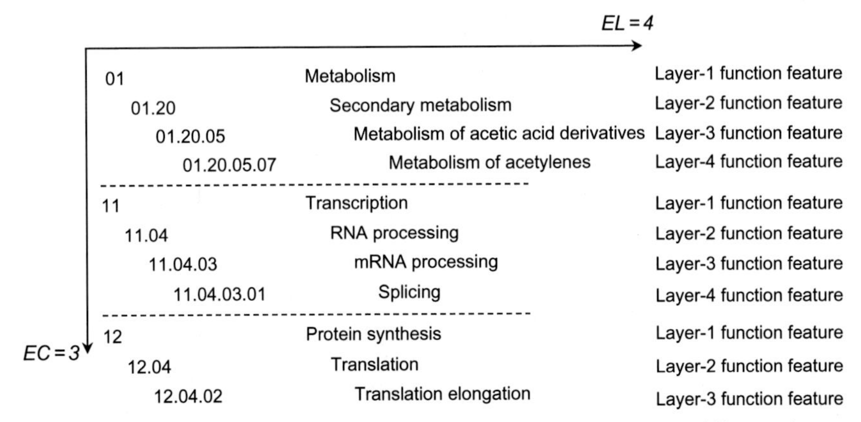

Fig. 3.1 Example of function feature selection of a cluster at different layers ($EC=3$, $EL=4$).

functions flexibly at different functional layers and in different functional categories.

Not all functions have the same number of functional layers. Two cases need to be considered when selecting the feature functions of a cluster. If the number of layers of a function is less than the number of expected layers (EL) and the number of expected layers is less than or equal to 5 ($EL \leq 5$), the function is excluded from the feature function selection at the expected layer. In the preceding example, if $EL=4$, then the function 12.04.02 is excluded from the layer-4 feature functions of the cluster. If the number of expected layers is 6 ($EL=6$), which means the real specific functions are expected to be predicted, then all functions participate in the feature function selection.

Cluster feature function weighting: Cluster feature function weight measures the importance of a selected feature function within a cluster. Therefore, it is an important local measurement of a function within the cluster. Suppose the number of selected feature functions of cluster c is n_c and the rank of function f in cluster c is k, then the weight $W_f(c)$ of function f within cluster c is defined as

$$W_f(c) = 1 - \frac{k-1}{n_c}. \tag{3.5}$$

For example, if f is ranked 1 ($k=1$, the highest rank), then $W_f(c)=1$; if f is ranked n_c ($k=n_c$, the last one), then $W_f(c)=1/n_c$. This weighting is only to distinguish the importance of selected feature functions within a cluster from computational perspectives. It does not have any biological meanings.

Interaction weight calculation between an unannotated protein and its interactive clusters: Interaction weight measures the impact of a cluster on the function prediction of an unannotated protein in the context of all clusters that interact with the unannotated protein. A cluster is considered as having an interaction with an unannotated protein if there is at least one interaction between the unannotated protein and any proteins in the cluster. Suppose the unannotated protein is u, the total number of clusters that interact with the protein u is n, and $m_u(c)$ is the number of interactions between the protein u and the cluster c. The interaction weight $P_u(c)$ between protein u and cluster c is defined as

$$P_u(c) = \frac{m_u(c)}{\sum_{i=1}^{n} m_u(i)}. \tag{3.6}$$

The higher interaction weight indicates that cluster c has more impact on function prediction of protein u.

Function scoring and prediction: Function prediction of an unannotated protein is based on the weights of feature functions in the clusters that interact with the unannotated protein, as well as the interaction weights between the unannotated protein and these clusters. A predicted function of the unannotated protein should be a feature function which has a high weight within a cluster that has a high interaction weight with the unannotated protein. With the feature function weight definition (3.5) and the interaction weight definition (3.6), the contribution $FW_{u,f}(c)$ of a feature function f from cluster c to the function prediction of the unannotated protein u is defined as

$$FW_{u,f}(c) = \begin{cases} P_u(c) \times W_f(c) & \text{if } f \in CF_c; \\ 0 & \text{if } f \notin CF_c, \end{cases} \tag{3.7}$$

where CF_c is the set of feature functions of cluster c. Fig. 3.2 gives an example of calculating the contribution of three feature functions (f_1, f_2, and f_3) from cluster 1 to the function prediction of unannotated protein u.

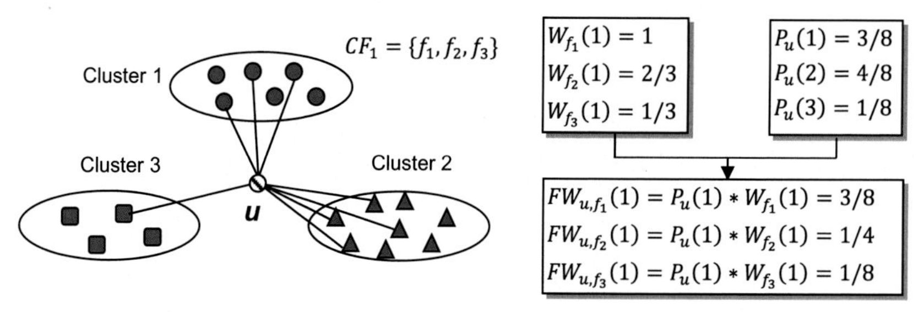

Fig. 3.2 Example of calculating the contribution of feature functions to function prediction.

Suppose the number of clusters that interact with the unannotated protein u is n, then the contribution of feature function f from all interactive clusters of u to the function prediction of u is defined as the following function score:

$$S_u(f) = \sum_{c=1}^{n} FW_{u,f}(c), \quad f \in \bigcup_{c=1}^{n} CF_c. \tag{3.8}$$

All the feature functions from the interactive clusters of the unannotated protein u are ranked in a descending order according to their function scores. The highly ranked feature functions (e.g., the first three) are selected as the predicted functions of the unannotated protein u. It can be seen from Fig. 3.2 that the final predicted functions may come from different clusters rather than from only one cluster. This feature makes the preceding algorithm different from other conventional clustering-based function prediction algorithms.

3.4 CLUSTER FEATURE FUNCTION SELECTION

It can be seen from the function prediction algorithm in Section 3.3 that cluster feature function selection is the basis of the prediction, which in turn has a great impact on the prediction quality. With the preceding feature function selection algorithm, the feature function selection at a functional layer depends on the feature function selection at the previous layer(s) or ancestral layer(s) of this layer. This feature function selection algorithm is suitable for the cases where the function prediction is to trace the most common functions at each functional layer. However, in some function prediction cases, it is expected that the prediction is made directly at a certain functional layer without considering the relationship with the ancestral layers. For these cases, another feature function selection algorithm, named Independent Selection (*IS*) in Zhu et al. (2014), can be a choice.

The *IS* algorithm directly selects feature functions at a specified functional layer from all functions in a cluster. For example, if the expected layer is layer-3, the algorithm directly extracts the first three layers of each function in the cluster and selects the most frequent 3-layer functions as the cluster feature functions at layer-3. Therefore, if the expected functional layer of the function prediction is six (i.e., the deepest layer in the FunCat annotation scheme), the *IS* selection algorithm becomes the one used by other existing non-layered clustering-based prediction methods. In other words, the *IS* algorithm is more general and supports layered function predictions.

Similar to the feature function selection in Section 3.3, the *IS* algorithm also first ranks functions according to the frequency of their occurrence within the cluster. However, the selection of feature function is based on a threshold, that is, only those functions with occurrence frequencies above a threshold are selected as the cluster feature functions. Specifically, let f_i be the occurrence frequency of a distinct function i in a cluster, then the threshold ε for feature function selection is calculated by

$$\varepsilon = \sum_{i=1}^{n} f_i w_i, \quad \text{with } w_i = f_i / \sum_{k=1}^{n} f_k \tag{3.9}$$

where n is the number of distinct functions within the cluster. The threshold ε is actually a weighted average of frequencies; therefore its value is within the range of minimum and maximum function occurrence frequencies in a cluster. For the layered cluster feature function selection, at different functional layers, the values of n, ε will be different. When selecting feature functions at a specified functional layer, if a function i satisfies the condition $f_i \geq \varepsilon$, it is then selected as a cluster feature function.

The *IS* algorithm of selecting cluster feature functions at a specified functional layer is depicted in the following pseudo-code. The definitions of the notations used in the pseudo code are listed here:

$i(l)$: The functional notation of function i at functional layer l in the FunCat scheme. For example, if function i is annotated as 18.02.01.02.03.0, then $i(3)$ is 18.02.01.

$f_{c_k}^{(i,l)}$: The occurrence frequency of $i(l)$ within cluster c_k.

$\varepsilon_{c_k}^{(l)}$:The threshold for feature function selection at functional layer l in cluster c_k.

$F_{c_k}^{l}$: The set of feature functions of cluster c_k at functional layer l.

$DF^{(l)}$: The set of distinct functions at functional layer l. For example, for three functions i = 18.02.01.02.03.0, j = 18.02.01.04.03.0, and k = 10.01.05.01.0.0, $DF^{(3)} = \{18.02.01, 10.01.05\}$.

Independent Selection (*IS*) Algorithm:

Input: Specified functional layer L ($1 \leq L \leq 6$);
Protein clusters $c_k(k = 1, \ldots, n)$. // n is the number of clusters.

Output: $F_{c_k}^{L}$ - The feature function sets of clusters $c_k(k = 1, \ldots, n)$ at specified layer L.

Algorithm:

For $k = 1$ to n

$\{F_{c_k}^{L} = \varnothing;\}$

For $k = 1$ to n

{

Construct $DF^{(L)}$ for all functions in cluster c_k;

For each function $i \in DF^{(L)}$, calculate $f_{c_k}^{(i,\,L)}$;

Calculate the selection threshold $\varepsilon_{c_k}^{(L)}$ according to(3.9);

For each function $i \in DF^{(L)}$

$\{$If $f_{c_k}^{(i,\,L)} \geq \varepsilon_{c_k}^{(L)}$ then $F_{c_k}^{L} = F_{c_k}^{L} \cup \{i\};\}$

$DF^{(L)} = \varnothing$;

}

3.5 DYNAMIC CLUSTERING-BASED FUNCTION PREDICTION

The existing clustering-based function prediction methods predict functions mainly from the final clusters generated by clustering methods. In fact, clustering proteins for function prediction is a dynamic process with which the possible predicted functions are grouped into the clusters that are related to the unannotated proteins. Although the existing clustering methods have been well developed, how to use these clustering methods to effectively predict functions is still a challenging issue. For example, if the *k*-means clustering method is used, the value of the clustering parameter *k* has an impact on the quality of clustering and, in turn, on the quality of the function prediction. Similarly, if a hierarchical clustering algorithm is used, the extent to which the clustering is conducted and the way of predicting functions from the clusters have a significant impact on the quality of cluster and prediction. On the other hand, protein interaction data sets usually contain a lot of false positive interactions (Deane et al., 2002) which might result in the irrelevant proteins being improperly clustered into the clusters from which functions are predicted. These issues indicate that predicting protein functions only from the final generated clusters may result in low prediction quality. These issues should be addressed and resolved when using clustering-based approaches to predict functions of unannotated proteins.

In this section, an innovative function prediction method is introduced. This method is based on dynamic clustering and addresses the preceding issues of existing clustering-based methods. Actually, in addition to considering the final generated clusters when predicting functions, this new method also traces the occurrences of functions across all other generated clusters that are relevant to the unannotated proteins during the recursive clustering processes, and selects those functions as the predicted functions that have a higher stability in this recursive clustering process. The idea behind this method is that, if a function always or in most cases accompanies the unannotated protein in the clusters in which the unannotated protein is located during the recursive clustering processes, this function is most likely to be a function of the unannotated protein. The prediction based on the recursive clustering can also effectively reduce the impact of noise information in the PPI network on the prediction results.

Since the function prediction is based on tracing function occurrences across all recursively generated clusters that are relevant to the unannotated protein, the unannotated protein is going to participate in this recursive clustering process. Therefore a new similarity or distance between proteins (including unannotated proteins) in a protein interaction network needs to be defined, based on which recursive protein clustering is conducted to predict functions of unannotated proteins.

It is known that interactions between proteins in a PPI data set can be modelled as a network, within which a node represents a protein and an edge between two nodes represents the interaction between two proteins. Based on this network model, the similarity between two proteins can be defined from the number of common interaction proteins they share. The more common interaction proteins the two proteins share, the higher the similarity between the two proteins. Specifically, suppose a and b are two proteins in a protein interaction network, N_a is the set of proteins directly interacting with a and N_b is the set of proteins directly interacting with b, then the similarity $sim(a, b)$ between the proteins a and b is defined as

$$sim(a,b) = \begin{cases} \dfrac{|N_a \cap N_b|}{|N_a \cup N_b|}, & \text{if } a \text{ and } b \text{ are not interacting} \\ \dfrac{|N_a \cap N_b| + 1}{|N_a \cup N_b|}, & \text{if } a \text{ and } b \text{ are interacting} \end{cases} \tag{3.10}$$

Fig. 3.3 illustrates the similarity calculation between two proteins a and b in two different cases. From the similarity defined by (3.10), the distance between any two proteins a and b in a PPI network is defined as

$$dis(a,b) = \begin{cases} \dfrac{1}{sim(a,b)} - 1, & \text{if } sim(a,b) \neq 0 \\ \dfrac{1}{\alpha}, & \text{if } sim(a,b) = 0 \end{cases} \tag{3.11}$$

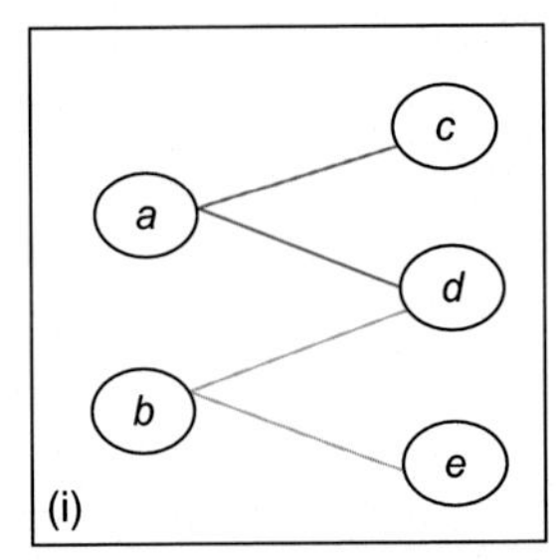

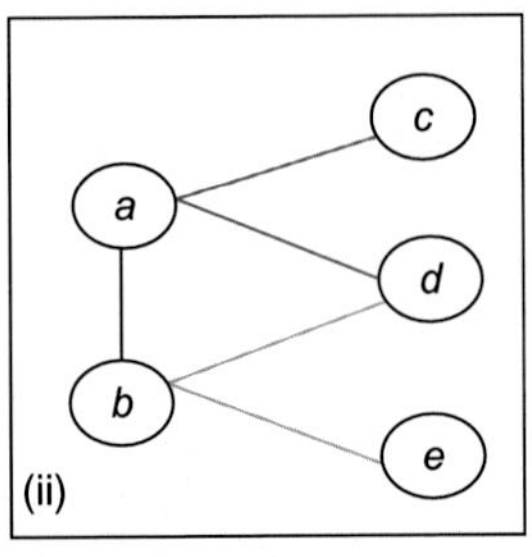

Fig. 3.3 Illustrations of similarity calculation between two proteins *a* and *b*. For case (i) where the proteins *a* and *b* are not interacting, $sim(a, b) = 1/3$. For case (ii) where the proteins *a* and *b* are interacting, $sim(a, b) = 2/5$.

where α is a constant which is small enough to make the distance value much bigger than the distance value when $sim(a, b) \neq 0$. The value of α can be set as $\alpha = 10^{-3}$ based on the experimental observations in Saini and Hou (2013).

With the similarity or distance definition between two proteins, theoretically any one of the existing clustering algorithms can be recursively used to cluster proteins in a protein interaction data set, and then the prediction algorithm that is described in details as follows is used to predict the functions of unannotated proteins from the recursively generated clusters. Specifically, the *k*-means algorithm (MacQueen, 1967) with $k = 2$ is selected for clustering operations in this method. Choosing $k = 2$ for the recursive *k*-means clustering enables the algorithm to include as many of the proteins that are functionally similar to the unannotated proteins as possible in the recursively generated clusters. Meanwhile, this recursive clustering procedure overcomes the limitations of the one-off *k*-means clustering method, where the value of the parameter *k* has a great impact on the prediction quality. On the other hand, function prediction of an unannotated protein depends on the context or domain it is located in. It has been observed that the functions of an unannotated protein are mainly located within its level-1 and level-2 neighbours (Chua et al., 2006). The level-1 neighbours of an unannotated protein are those proteins in the PPI network that interact directly with the unannotated protein, while the level-2 neighbours of an unannotated protein are those proteins in the PPI network that do not directly interact with the unannotated protein but do interact directly with the level-1 neighbours of the unannotated protein. Based on this observation, for an unannotated protein in this prediction method, its level-1 and level-2 neighbours are selected to form a sub-data set on which recursive clustering is conducted and functions are predicted. This sub-data set is named the *clustering domain* of an unannotated protein and is denoted as C_0.

Function prediction for an unannotated protein begins with recursively or progressively clustering the proteins in the clustering domain. The recursive clustering process focuses only on the clusters that contain the unannotated protein. This recursive clustering process keeps going until the unannotated protein is separated from other proteins in the clustering domain. Meanwhile, this recursive clustering process generates a size-decreasing chain of clusters that contain the unannotated protein. Denote these clusters as $C_0, C_1, \ldots, C_n$ which will be used to predict functions of the unannotated protein. The last cluster C_n, which is called the *probe cluster*, satisfies the condition that if the proteins in the cluster are further clustered, the unannotated protein will be separated from the rest (i.e., one subcluster of C_n contains only the unannotated protein). Fig. 3.4 graphically describes this recursive clustering process and the selection of the clusters for function prediction.

Based on the clusters generated from the recursive clustering operations, the functions of the unannotated protein are predicted. Specifically, the candidates of predicted functions are chosen from those functions that are possessed by at least one protein in the probe cluster C_n. The prediction is made based upon how the proteins that have these candidate functions are split during the recursive clustering operations (i.e., the recursive binary splitting—details are presented in the prediction algorithm description) as well as the occurrence traces of these candidate functions in the recursive clustering operations. To trace the occurrences of these functions during the clustering operations, for a function a in the probe cluster C_n, its

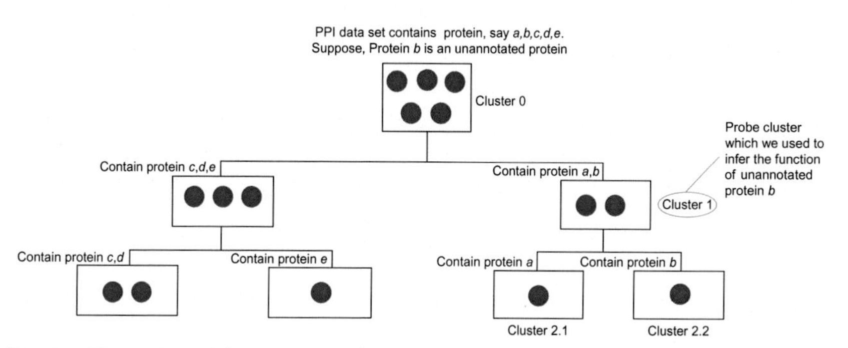

Fig. 3.4 Illustration of the recursive clustering process and the selection of the clusters for function prediction. Suppose protein *b* is an unannotated protein and is the only protein in Cluster 2.2 after the recursive clustering operations. This means Cluster 1 becomes the *probe cluster*, and Cluster 0 and Cluster 1 (i.e., C_0, C_1 for this example) are selected for predicting functions of the unannotated protein *b* (Saini and Hou, 2013).

Table 3.1 An example of tracing function frequency across clusters

Protein function	Function frequency in C_0	Function frequency in C_1	Function frequency in C_2	Function frequency in C_3
f_1	60	43	1	1
f_2	21	14	1	1
f_3	**89**	65	3	1
f_4	81	76	3	1

Note: This function tracing table shows the frequencies of the annotated protein functions (f_1, f_2, f_3, f_4) across the all clusters C_0, C_1, C_2, C_3 generated from recursive clustering operations. The traced four functions come from the *probe cluster* C_3.

occurrence frequency in cluster C_i $(0 \leq i \leq n)$ is defined as f_i^a, which is the number of proteins in C_i that possess the function a. Table 3.1 provides an example of tracing four functions across the generated clusters C_0, C_1, C_2, C_3 in terms of frequency.

The basic idea of this prediction algorithm is to select those functions in the *probe cluster* that are, on average, stable in terms of occurrence frequency across all generated clusters $C_0, C_1, \ldots, C_n$, and have a high probability of becoming the predicted functions. To this end, new metrics are defined to measure the average stability of a function across generated clusters and the probability of a function being identified as a predicted function.

The average stability (AS) of a function a in terms of occurrence frequency is defined as

$$AS^a = \frac{1}{n} \sum_{k=1}^{n} \frac{f_{k-1}^a - f_k^a}{f_{k-1}^a} \tag{3.12}$$

where n is the number of generated clusters. It can be seen from this average stability definition that the smaller the value AS^a, the more stable the function a. Generally, if a function is more stable across all generated clusters, it is more likely to be a predicted function. Therefore the probability of a function a being identified as a predicted function is accordingly defined as $(1 - AS^a)$. However, if a function is too stable (e.g., almost all proteins have this function), it may have no features that are special to the unannotated protein. On the other hand, if a function is too unstable (i.e., the AS value is too high), its probability of being identified as a predicted function is low, but it may have specific features that are special to the unannotated protein. The prediction method is to select those proteins that are stable as well as have a high probability of being identified as a predicted function. To incorporate these two factors into function prediction, the *fScore* of function a is defined as

$$fScore^a = AS^a \times (1 - AS^a) \tag{3.13}$$

Accordingly, the prediction algorithm is to select those functions with the higher *fScore* values as the predicted functions. To define a selection threshold, some concepts are defined as follows.

Let $D_F = \{\text{functions in the probe cluster } C_n\}$. The threshold value (*TValue*) is defined as

$$TValue = \max_{a \in D_F} \left(f_0^a\right)$$

The function in C_n that has the *TValue* is called *Tfunction*. For example, in Table 3.1, *TValue* = 89 and f_3 is the *Tfunction*. It can be seen that the *Tfunction* has the highest frequency in the clustering domain with regard to the functions in C_n and therefore is most likely to be a predicted function. The *Tfunction* is to be used as a preliminary threshold for selecting predicted functions. Based on these definitions, the protein function prediction algorithm is depicted as the pseudo-code in Fig. 3.5. It can be seen from the algorithm that the prediction is to select predicted functions globally over all function traces (Step 1) and locally over the last two clusters (Step 2). The descriptions of these steps are as follows.

The basic idea behind Step 1 is that the *Tfunction*, which possesses the *Tvalue*, is most likely to be a predicted function, and if a function has a higher *fScore* than the *fScore* of the *Tfunction*, that function is also likely to be a predicted function. For the case where *fScore* of a function is zero, which means the function is consistently stable across all clusters, the function is also selected as a predicted function. The second part of Step 1 determines whether or not the *Tfunction* should be a predicted function by checking the frequency of the *Tfunction* in the second last cluster, that is, $f_{n-1}^{Tfunction}$. If in the C_{n-1}, the *Tfunction* has frequency 1 (which is the lowest frequency in a cluster), there is little chance the *Tfunction* is a predicted function. The algorithm does not consider the probe cluster C_n, because there may be a case where it has only two proteins in it; one is annotated, and another is unannotated. In this case, the *Tfunction* always has the frequency 1 (which is the highest, as well as the lowest, frequency) in C_n, so it is uncertain whether the *Tfunction* should be a predicted function. But if the *Tfunction* has the frequency 1 in C_{n-1}, then it definitely has the frequency 1 in C_n. In other words, the *Tfunction* has the lowest frequency in the last two clusters. Since the last two clusters, C_{n-1} and C_n, are the most relevant clusters to the unannotated protein, the *Tfunction* should not be a predicted function in this case.

Step 2 of the algorithm is to select those functions in the probe cluster C_n that show the frequency significance in the last two clusters C_n and C_{n-1} but have *fScores* lower than the *fScore* of the *Tfunction*. This is because in Step 1

```
Initialize P_f = ∅   // P_f is the set of predicted functions

//Step 1: select predicted functions global ly over all function frequency traces
For any function a in the probe cluster C_n
{
        if fScore^a ≥ fScore^Tfunction
        then P_f = P_f ∪ { a };

        if fScore^a = 0 and f_0^a ≠ |C_0| − 1   // |C_0| is the number of proteins in cluster C_0
        then P_f = P_f ∪ { a };
}
For the Tfunction in the probe cluster C_n
{
        If f_{n−1}^Tfunction ≠ 1
        then P_f = P_f ∪ {Tfunction}
}

//Step 2: select predicted functions locally over the last two clusters
For any function a in the probe cluster C_n, compare its function frequencies in the C_n and C_{n−1}
{
        //Step 2A
        check f_n^a and f_{n−1}^a
        {
                If f_n^a = |C_n| − 1 and f_{n−1}^a = |C_{n−1}| − 1
                then P_f = P_f ∪ { a }
        }
        //Step 2B
        compare frequencies of the function a and the Tfunction
        {
                if f_n^a = f_n^Tfunction, f_{n−1}^a = f_{n−1}^Tfunction and f_{n−1}^Tfunction ≠ 1
                then P_f = P_f ∪ { a }
        }
        //Step 2C
        check the probe cluster C_n
        {

                If there exist functions a, b,..., m in C_n such that

                        f_n^a ≥ f_n^Tfunction, f_n^b ≥ f_n^Tfunction, ..., f_n^m ≥ f_n^Tfunction

                    and f_n^a ≠ 1, f_n^b ≠ 1, .... f_n^m ≠ 1
                then compare fScore^a, fScore^b, ..., fScore^m
                {
                        If fScore^a ≥ fScore^b ≥ ··· ≥ fScore^m
                        then P_f = P_f ∪ { a }
                }
        }
}
Return P_f
```

Fig. 3.5 Pseudo-code of the function prediction algorithm.

some functions that are not the real ones of the unannotated protein might have higher *fScores*, while the real functions might have relatively low *fScores*, because of the noisy PPI information. Since the last two clusters C_n and C_{n-1} are the most relevant clusters to the unannotated protein, Step 2 makes it possible that some real functions of the unannotated protein are not overwhelmed by the noisy information in the original PPI data in terms of the frequency significance. This step considers three possible cases locally over the clusters C_n and C_{n-1}, which correspond to the following three sub-steps:

Step 2A: This sub-step selects those functions as the predicted functions that consistently have a maximum frequency over the clusters C_n and C_{n-1}, because a function with the maximum frequency in the last two clusters is more likely to be closely related to the unannotated protein.

Step 2B: This sub-step selects those functions that do not have a maximum frequency over the last two clusters, but do have the same frequency pattern as the *Tfunction*. Since the *Tfunction* with $f_{n-1}^{Tfunction} \neq 1$ is a predicted function (see Step 1), a function that consistently shows the same frequency pattern as the *Tfunction* over the last two clusters is also likely to be a predicted function.

Step 2C: This sub-step selects those functions whose frequencies in the last cluster C_n are higher than or equal to the frequency of the *Tfunction*, but are not selected as the predicted functions in Step 1 because of their *fScores* being lower than that of the *Tfunction*. These functions appear in more proteins than the *Tfunction* in the most relevant cluster C_n. Therefore they also have a great possibility of being the predicted functions. However, considering the possible noisy PPI information, as well as both the frequency and *fScore* of each function, only one of these functions that has the highest *fScore* is selected as the predicted function. This sub-step keeps a balance between selecting more real functions and reducing the impact of possible noisy information in the prediction.

3.6 DISCUSSIONS

It can be seen from the two function prediction methods in this chapter that their ways of treating the unannotated protein in the clustering process are different. In the dynamic clustering-based function prediction method (Section 3.5), the unannotated protein is involved in the clustering operations. This is because the function prediction is based on selecting the recursively generated clusters in which the unannotated protein is located and tracing the function occurrence frequencies in these clusters. Accordingly, the protein similarity is defined based on the topological information of a PPI interaction network.

In the layered protein function prediction method (Section 3.3), the unannotated protein is not involved in the clustering operations. This treatment reduces the impact of semantic uncertainty between the unannotated protein and its interaction partners on the clustering results and, in turn, on the prediction results. This is because the functions of the unannotated protein are unknown before clustering, any ways to assign initial functions to the unannotated protein or any assumed semantic relationship between the unannotated protein and its interaction partners will unavoidably bring noise information into the clustering and prediction procedure.

Furthermore, not including the unannotated protein in the clustering process makes it possible to select similar, as well as complementary, functions from different clusters as the candidates from which the functions of the unannotated protein are predicted. This approach can more reasonably reflect the biological realities of protein interactions; that is, proteins interact with each other to implement a biological process by providing similar and complementary functions. So it is most likely that the functions of the unannotated protein contain some functions that are similar, as well as complementary, to the functions of its interaction partners. This approach of treating the unannotated protein in the clustering process overcomes the limitation of many other prediction methods that focus on finding those similar functions as the predicted functions.

The layered function prediction method in Section 3.3 makes it more flexible to predict and understand protein functional characteristics at different functional layers, rather than just the specific functions. The prediction results at expected layer(s) also provide multiple guidelines in applications such as biological experiment design. This is one of the features that differentiates this method from other clustering-based prediction methods.

The ranking and selection of cluster feature functions are critical to the prediction quality of the layered function prediction method. The ranking and selection algorithm in Section 3.3 selects only the highest ranked function as the feature function at each layer. Actually, this algorithm could be extended to select more cluster feature function candidates at each functional layer, such as selecting the first, second or more highly ranked functions as the feature functions, or selecting feature functions from the ranked functions according to a threshold. The Cascaded Selection (CS) algorithm in (Zhu et al., 2014) defines a selection threshold and extends this algorithm to select more feature functions at each functional layer. However, this extension is a two-edged sword because, on one hand, the prediction domain will be wider, which possibly covers the real functions of the unannotated proteins; on the other hand, noise functions might be brought into the prediction domain, which distorts the prediction results. More investigations are needed to find a balance point between the prediction coverage and precision when selecting feature functions of a cluster.

Although the layered function prediction method in this chapter relies on the FunCat-based function similarity (Section 3.2), other function similarities can also be adapted by this method as long as the similarity definition is based on a hierarchical function annotation scheme such as GO. In the FunCat-based function similarity definition, the deepest functional layer each function has is fixed, that is, layer 6. So the similarity definition in Section 3.2 is relatively easy and direct. If another hierarchical function annotation scheme, GO for example, is used, the deepest functional layer of different functions will be different and is not fixed at 6. So if the idea in the FunCat-based similarity definition is used to define new similarities, it will be necessary to set some variable parameters in the new similarity definitions. This might be another further research topic based on the methods of this chapter.

REFERENCES

Adamcsek, B., Palla, G., Farkas, I.J., Derényi, I., Vicsek, T., 2006. CFinder: locating cliques and overlapping modules in biological networks. Bioinformatics 22 (8), 1021–1023.

Arnau, V., Mars, S., Marin, I., 2005. Iterative cluster analysis of protein interaction data. Bioinformatics 21, 364–378.

Asur, S., Ucar, D., Parthasarathy, S., 2007. An ensemble framework for clustering protein-protein interaction networks. Bioinformatics 23, i29–i40.

Brohée, S., van Helden, J., 2006. Evaluation of clustering algorithms for protein-protein interaction networks. BMC Bioinf. 7, 48.

Brun, C., Chevenet, F., Martin, D., Wojcik, J., Guenoche, A., Jacq, B., 2003. Functional classification of proteins for the prediction of cellular function from a protein–protein interaction network. Genome Biol. 5, R6.

Brun, C., Herrmann, C., Guenoche, A., 2004. Clustering proteins from interaction networks for the prediction of cellular functions. BMC Bioinf. 5 (1), 95.

Chen, J., Yuan, B., 2006. Detecting functional modules in the yeast protein-protein interaction network. Bioinformatics 18, 2283–2290.

Chua, H.N., Sung, W.K., Wong, L., 2006. Exploiting indirect neighbours and topological weight to predict protein function from protein-protein interactions. Bioinformatics 22 (13), 1623–1630.

de Hoon, M.J.L., Imoto, S., Nolan, J., Miyano, S., 2004. Open source clustering software. Bioinformatics 20 (9), 1453–1454.

Deane, C.M., Salwinski, L., Xenarios, I., Eisenberg, D., 2002. Protein interactions: two methods for assessment of the reliability of high-throughput observations. Mol. Cell. Proteomics 1 (5), 349–356.

Dunn, R., Dudbridge, F., Sanderson, C., 2005. The use of edge-betweenness clustering to investigate biological function in pins. BCM Bioinf. 6, 39.

Enright, A.J., Dongen, S.V., Ouzounis, C.A., 2002. An efficient algorithm for large-scale detection of protein families. Nucleic Acids Res. 30 (7), 1575–1584.

Fortunato, S., 2010. Community detection in graphs. Phys. Rep. 486, 75–174.

GO Consortium, 2016. Ontology structure. http://geneontology.org/page/ontology-structure (accessed 11.03.16).

King, A.D., Przulj, N., Jurisica, I., 2004. Protein complex prediction via cost-based clustering. Bioinformatics 20, 3013–3020.

Luo, F., Yang, Y., Chen, C.F., Chang, R., Zhou, J., Scheuermann, R.H., 2007. Modular organization of protein interaction networks. Bioinformatics 23 (2), 207–214.

Maciag, K., Altschuler, S.J., Slack, M.D., Krogan, N.J., Emili, A., Greenblatt, J.F., Maniatis, T., Wu, L.F., 2006. Systems-level analyses identify extensive coupling among gene expression machines. Mol. Syst. Biol. 2, 3.

MacQueen, J.B., 1967. Some methods for classification and analysis of multivariate observations. In: Proceedings of the Fifth Berkeley Symposium on Mathematical Statistics and Probability, vol. 1. Statistical Laboratory of the University of California, Berkeley, pp. 281–297.

Pereira-Leal, J.B., Enright, A.J., Ouzounis, C.A., 2004. Detection of functional modules from protein interaction networks. Proteins 54, 49–57.

Przulj, N., Wigle, D.A., Jurisica, I., 2004. Functional topology in a network of protein interactions. Bioinformatics 20, 340–348.

Ruepp, A., Zollner, A., Maier, D., Albermann, K., Hani, J., Mokrejs, M., Tetko, I., Güldener, U., Mannhaupt, G., Münsterkötter, M., 2004. The FunCat, a functional annotation scheme for systematic classification of proteins from whole genomes. Nucleic Acids Res. 32 (18), 5539–5545.

Saini, A., Hou, J., 2013. Progressive clustering based method for protein function prediction. Bull. Math. Biol. 75 (2), 331–350.

Samanta, M.P., Liang, S., 2003. Predicting protein functions from redundancies in large-scale protein interaction networks. Proc. Natl. Acad. Sci. U. S. A. 100 (22), 12579–12583.

Trivodaliev, K., Bogojeska, A., Kocarev, L., 2014. Exploring function prediction in protein interaction networks via clustering methods. PLoS ONE 9(6), e99755.

Zhu, W., Hou, J., Chen, Y.-P.P., 2010. Semantic and layered protein function prediction from PPI networks. J. Theor. Biol. 267, 129–136.

Zhu, W., Hou, J., Chen, Y.-P.P., 2014. Select cluster features for better layered protein function prediction. Curr. Bioinforma. 9 (3), 306–314.

CHAPTER 4

Iterative Approaches of Protein Function Prediction

The quality of function prediction from protein interactions relies greatly on whether we can extract and exploit proper information (semantic or non-semantic) from protein associations that really reflect the intrinsic functional nature of protein relationships. The majority of similarity-based prediction approaches regard the prediction as a mono-directed and one-off procedure; that is, the available functions of annotated proteins in the prediction domain (e.g., the neighbour proteins of an unannotated protein), as well as their mutual relationships, determine what functions an unannotated protein should have (mono-directed prediction), and once the functions of unannotated proteins are predicted, the prediction is finished (one-off prediction). During the prediction process, the unannotated proteins are passive and have no functional impact on the relationship between proteins.

As a matter of fact, in real biological processes, proteins have high mobility and dynamically interplay to produce a framework which is ever-changing but overall stable (Misteli, 2001). The proteins exchange their biological information and share functions in a dynamic, rather than a static and mono-directed, circumstance. This fact indicates that the dynamic and mutual interaction between pair-wise proteins, including unannotated proteins although their functions are yet to be annotated or predicted, is the reality and nature of biological processes, and this dynamic feature of protein interaction should be taken into account when predicting functions.

This chapter introduces three iterative methods for protein function prediction: local, semi-local, and global iterative methods. These methods address the preceding issues of function prediction with respect to different prediction domains and take generally similar but technically different ways to iteratively predict protein functions. The functional association in terms of semantic similarity between proteins is integrated into the iterative prediction procedures which simulate the functional dynamic situations of protein interaction. The algorithms of these methods make the proteins and functions endorse each other through the iterative procedures, as well as

New Approaches of Protein Function Prediction from Protein Interaction Networks
http://dx.doi.org/10.1016/B978-0-12-809814-1.00004-2
© 2017 Elsevier Ltd.
All rights reserved.

select the most semantically important functions as the predicted functions of unannotated proteins.

4.1 LOCAL ITERATIVE FUNCTION PREDICTION METHOD

This prediction method is based on dynamic functional voting from the neighbour proteins of the unannotated protein. The functions that obtain higher votes or voting scores from the neighbour proteins are the predicted functions of the unannotated protein. The votes a function obtains depend not only on the importance of the neighbour proteins that have the function but also on the importance of the function among all available functions possessed by the neighbour proteins. These two kinds of importance are measured in terms of protein semantic similarity (i.e., the protein similarity defined by function similarities) and function similarity, respectively.

Suppose the semantic similarity between two proteins p and p' has already been defined as $sim(p,p')$, and the similarity between two functions f and f' has already been defined as well as $fsim(f,f')$. Let $N(p)$ be the set of neighbour proteins of the protein p, $F(p)$ be the set of functions protein p has, and $NF(p)$ be the set of functions the protein p's neighbour proteins have; that is, $NF(p) = \cup_{p' \in N(p)} F(p')$. For an unannotated protein p, the scores a function $f \in NF(p)$ can obtain from functional voting of the p's neighbour proteins are defined as

$$score(p,f) = \sum_{p' \in N(p)} \left[sim(p,p') \times \sum_{f' \in F(p')} fsim(f,f') \right] \qquad (4.1)$$

For this prediction method, the neighbour protein set $N(p)$ of the unannotated protein p does not contain any other unannotated proteins. For the cases where $N(p)$ also contains unannotated proteins whose functions are yet to be predicted, the global iterative prediction method in Section 4.3 is the choice.

It can be seen from Eq. (4.1) that the majority rule method proposed by Schwikowski et al. (2000) is actually a special case of this prediction method if $sim(p,p')$ in Eq. (4.1) is set to 1 (i.e., $sim(p,p') = 1$), the component $\sum_{f' \in F(p')} fsim(f,f')$ in Eq. (4.1) is replaced by a simple indicator function (i.e., if p' has function f then the value is 1, otherwise 0) without considering function associations, and $N(p)$ consists of only the annotated proteins that directly interact with the unannotated protein p, that is, the level-1 neighbour proteins of p.

From Eq. (4.1), the scores a function f obtains from the neighbour protein voting are determined by two factors: the importance of each neighbour protein to the unannotated protein p (i.e., $sim(p, p')$ in Eq. 4.1), and the importance of function f in each neighbour protein (i.e., $\sum_{f' \in F(p')} fsim(f, f')$ in Eq. 4.1). For example, functions that are possessed by the most important neighbour proteins and are the most important among neighbour functions will achieve the highest scores, and therefore are most likely to be the predicted functions of the unannotated protein p. It is obvious a function possessed by the less important neighbour proteins but very important among neighbour functions, or possessed by very important neighbour proteins but less important among neighbour functions is also likely to obtain higher scores and to be a predicted function.

For a given unannotated protein p and its neighbour proteins, the importance of a function f in each neighbour protein, and in turn among neighbour functions (i.e., $\sum_{f' \in F(p')} fsim(f, f')$ in Eq. 4.1), is fixed when the prediction domain $N(p)$ is formed. Therefore the variation of the score depends on the variation of semantic similarities between the unannotated protein p and its neighbour proteins (i.e., $sim(p, p')$ in Eq. 4.1). However, since the functions of the unannotated protein p are unknown and yet to be predicted, the semantic similarity $sim(p, p')$ is therefore uncertain at the beginning of the prediction. This prediction method is to iteratively update the semantic similarities $sim(p, p')$ between the unannotated protein and its neighbour proteins and, in turn, to iteratively update function scores until they no longer change or achieve a stable status. When the iteration is finished, the functions with the higher scores are then selected as the predicted functions. To this end and also to reflect the dynamic features of protein interaction in this iterative semantic similarity updating process, a semantic similarity between proteins defined from their functional similarities is necessary so that the important neighbour proteins and important functions can endorse each other through the iterative updating by Eq. (4.1) and the functions that are functionally important to the unannotated protein can be selected as the predicted functions.

The function similarity in this method is defined based on the FunCat function annotation scheme developed by the Munich Information Centre for Protein Sequences (MIPS) (Ruepp et al., 2004). The FunCat scheme allows a protein function to be expressed numerically by up to six functional layers. A digital number at each layer defines a specific function category. The deeper a function's layer achieves, the more specific the function is.

Particularly, the similarity of two protein functions is determined by the number of common layers two functions share; that is, the more layers two functions share, the more similar two functions are. The details of the function similarity definition are as follows (Zhu et al., 2012; Hou et al., 2013).

For two given functions f and f' annotated in FunCat digital layer format, $l(f,f') \stackrel{\text{def}}{=}$ the number of common sequent layers the f and f' share from the first layer. It is obvious $0 \le l(f,f') \le 6$. For example, suppose f=10.01.05.03.01.0 and f'=10.01.03.0.0.0; these two functions share the first two layers 10.01 and therefore $l(f,f') = 2$. Based on this definition, the similarity $fsim(f,f')$ between two functions f and f' is defined as

$$fsim(f,f') = \sum_{i=1}^{l(f,f')} i^2 / \sum_{j=1}^{6} j^2 \tag{4.2}$$

If $l(f,f') = 0$, then $fsim(f,f') = 0$. With the function similarity definition in Eq. (4.2), the semantic similarity $sim(p_1, p_2)$ between two proteins p_1 and p_2 is defined as

$$sim(p_1, p_2) = \sum_{f \in F(p_1)} \sum_{f' \in F(p_2)} fsim(f,f') / \sum_{f \in F(p_1)} \sum_{f' \in F(p_2)} w(f,f') \tag{4.3}$$

where

$$w(f,f') = 2 - l(f,f')/6$$

This protein similarity definition is similar to the one defined in Chapter 3, but is slightly different and can more precisely measure the similarity between proteins.

To calculate the function scores in Eq. (4.1) with respect to the unannotated protein p, two questions need to be answered: one is how to choose the neighbour proteins $N(p)$ from which the functions are predicted; another is how to calculate the semantic similarities between the unannotated protein p and its neighbour proteins. Regarding the first question, this method chooses level-1 and level-2 neighbour proteins of the unannotated protein p to form the neighbour protein set $N(p)$, which is also known as the prediction domain of p. Level-1 neighbour proteins are those that directly interact with the unannotated protein p, while level-2 neighbour proteins are those that directly interact with the level-1 neighbour proteins of p but do not directly interact with the unannotated protein p. This neighbour protein set $N(p)$ construction is based on the work in Chua et al. (2006), which observed that in most cases, level-1 and level-2 neighbour proteins contain major functions of the protein p.

Regarding the second question, the issue is that functions of protein p are unknown while the semantic protein similarity $sim(p,p')$ is calculated from protein function similarities. Therefore, to kick off the prediction, the initial functions are needed to be assigned to the unannotated protein p. This function initiation is done by selecting the level-1 neighbour proteins of p and setting $sim(p, p') = 1$ in Eq. (4.1) to calculate the scores of level-1 functions. The level-1 functions are ranked according to their scores in descending order. Then the average number of functions each level-1 neighbour protein has is set as the cut-off rate for selecting top ranked level-1 functions to initialize the functions of the unannotated protein p. For example, if on average each level-1 neighbour protein has three functions, then the functions with the first three highest scores are selected as the initial functions of p.

With the initial functions assigned to the unannotated protein p, the calculation of the semantic similarities between the unannotated protein p and its neighbour proteins in Eq. (4.1) can be conducted. At this stage, the neighbour protein set $N(p)$ consists of level-1 and level-2 neighbour proteins rather than just level-1 proteins. For a level-2 neighbour protein, its similarity with the unannotated protein p is calculated a little bit differently than Eq. (4.3). In fact, suppose a level-2 protein is p_2, then the similarity between p and p_2 is calculated as

$$sim(p, p_2) = \max\left[sim(p, p_2), \max_{p_1 \in N1(p_2)} (sim(p, p_1) \times sim(p_1, p_2))\right] \quad (4.4)$$

where $N1(p_2) = \{p_1 | p_1$ is the level-1 neighbour of p and directly interacts with $p_2\}$.

With the issues of calculating the function score in Eq. (4.1) being resolved as previously described, function prediction can be conducted dynamically based on Eq. (4.1). In fact, with the initial functions being assigned to the unannotated protein p, the set $N(p)$ in Eq. (4.1) now consists of the level-1 and level-2 neighbour proteins of p, the scores of all functions in the neighbour are calculated and the functions are ranked according to their scores. The functions with higher scores are selected as the first-round predicted functions. The number of selected functions depends on the cut-off rate in the prediction. At this prediction stage, the cut-off rate is defined as the *average number* of functions each protein has in the neighbour set $N(p)$. If the current-round prediction results are the same as the previous ones, that is, the prediction results do not change anymore, the prediction operation is then stopped and the current-round prediction results are the final predicted functions of the unannotated protein (for the first round of prediction, its previous-round prediction results are those initial functions of the unannotated protein). Otherwise the current-round predicted functions are assigned to the unannotated protein, and the prediction based on Eq. (4.1) is conducted again until the results no longer change. This iterative prediction algorithm is depicted as a diagram in Fig. 4.1.

This iterative prediction algorithm is convergent. Actually, the iterative prediction procedure makes the protein similarities and function scores endorse each other iteratively by Eq. (4.1). In other words, if a function with the highest functional score is selected as a predicted function after the first round of iteration, it will be kept as a predicted function in the following iteration rounds as well, and each iteration round will select functions with the new highest functional score while the previously selected functions are still kept. This means that after finite rounds of iteration, the prediction

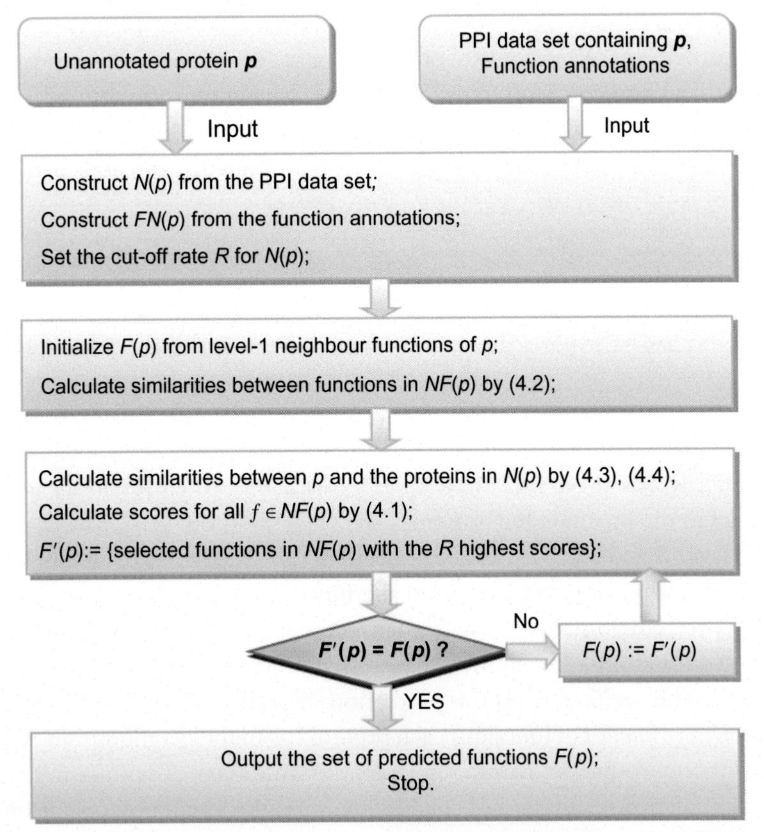

Fig. 4.1 Diagram of the iterative function prediction algorithm (Hou et al., 2013).

results will no longer change or the algorithm is convergent. The experiments on real protein interaction data sets (Hou et al., 2013) also demonstrated the convergence of the algorithm.

4.2 SEMI-LOCAL ITERATIVE FUNCTION PREDICTION METHOD

The iterative prediction methods, like the one in Section 4.2 and others to be introduced in this and the next section, are to simulate the dynamic process of protein interactions in terms of protein and function similarities when predicting protein functions. It refers to three major aspects: the selection of prediction domain, function (and in turn protein) similarity definition and prediction algorithm. Regarding the iterative prediction method in Section 4.1, it selects level-1 and level-2 neighbour proteins of an unannotated protein as the prediction domain, defines FunCat-based function and protein similarities and uses the function voting scores iteratively to predict functions of the unannotated protein. However, the iterative function voting is conducted only within the local prediction domain; no global information about functions in the whole interaction data set is taken into consideration. This section introduces another iterative method which selects the level-1 neighbour proteins of an unannotated protein as the prediction domain, defines function and protein similarities based on the Gene Ontology (GO) annotation scheme (GO Consortium, 2016) and takes into account the local as well as the global semantic influence of functions when predicting functions from the iterative function voting. This method could more reasonably count the contribution of available functions to the prediction results when the interaction data set is in good quality.

The iterative prediction of this method starts with assigning initial predicted functions to the unannotated protein and then calculates the initial similarities between the unannotated protein and its neighbour proteins. With these initial similarities, a *k*-nearest neighbour (*kNN*)-based prediction method is applied to get the new predicted functions for the unannotated protein. Replacing the initial/old predicted functions of the unannotated protein, the new predicted functions are then used to recalculate the similarities between the unannotated protein and its neighbour proteins for the next round of prediction. This prediction process is repeated until the similarities between the unannotated protein and its neighbour proteins reach a stable state, which represents a dynamic stable status among the protein interactions in terms of similarities.

The idea behind this prediction algorithm is to iteratively count the contribution of the available functions in the neighbours of the unannotated protein to the final determination of predicted functions. The contribution of a function to the prediction is primarily dependent on the number of neighbour proteins that have the function and the similarities between the unannotated protein and these neighbour proteins. Furthermore, the similarities between the functions in the neighbour, as well as the global and local influence of the functions, are also taken into consideration in the prediction. The details of this iterative prediction algorithm are presented as follows. Since the basis for this algorithm is the proper definitions of the protein similarity and function similarity which guarantee the prediction being conducted iteratively, the new GO-based protein and function similarities are introduced first as follows, followed by the prediction algorithm.

Suppose the unannotated protein is p, the neighbour proteins of p are denoted as a set $N(p)$. The neighbour proteins of a protein p are those that have direct and/or indirect interactions with p in the protein-protein interaction (PPI) network. In this method only those proteins that have direct interactions with p are selected as the neighbour proteins of p, that is, the level-1 neighbours. Denote the functions of a protein p' as a set $F(p')$, the functions of all neighbour proteins of p as another set $FN(p) = \cup_{p' \in N(p)} F(p')$, and assume all available protein functions in the interaction data set are annotated by the GO terms.

For any two proteins p and p', suppose the size of the set $F(p)$ is m (i.e., the number of functions in $F(p)$) and the size of the set $F(p')$ is n. The similarity $sim(p, p')$ between two proteins p and p' is defined as

$$sim(p, p') = \frac{1}{\max(m, n)} \sum_{f \in F(p)} \sum_{f' \in F(p')} \delta_{f,f'} \tag{4.5}$$

where $\delta_{f,f'}$ is an indicator function; that is, if f and f' are the same, its value is 1, otherwise 0.

Any two functions f and f', they can be represented as two vectors $\vec{f}$ and $\vec{f'}$ whose element values indicate the occurrences of the GO terms that annotate the functions. If the number of terms/notations in GO is t, the dimension of each function vector $\vec{f}$ is also t. Since GO is a directed acyclic graph in which a GO term may have multiple parent GO terms, all parent terms of a GO term are also called the ancestors of the term. If a function is annotated by a GO term, it is also annotated by the ancestors of the GO term. Therefore the vector element values at the index positions that correspond to these ancestors are set to 1, otherwise set to 0. For example, suppose there are five GO terms for functional annotation (just for demonstration only), a function f is annotated by the fourth term whose ancestors are the second and third terms, and another function f' is annotated by the fifth term whose ancestors are the third and fourth terms, then these two functions f and f' can be represented as two vectors $\vec{f} = (0, 1, 1, 1, 0)$ and $\vec{f'} = (0, 0, 1, 1, 1)$, respectively. The similarity $fsim(f, f')$ between two functions f and f' is defined as

$$fsim(f, f') = \vec{f} \cdot \vec{f'} / \left\| \vec{f} \right\| \cdot \left\| \vec{f'} \right\| \tag{4.6}$$

where $\vec{f} \cdot \vec{f'}$ is the dot product of two vectors and $\left\| \vec{f} \right\|$ is the norm of the vector $\vec{f}$. It can be seen that the similarity between two functions is their cosine similarity whose value is within the range of $0 \le fsim(f, f') \le 1$. For the preceding two example function vectors $\vec{f} = (0, 1, 1, 1, 0)$ and $\vec{f'} = (0, 0, 1, 1, 1)$ for instance, $\vec{f} \cdot \vec{f'} = 2$, $\left\| \vec{f} \right\| = \left\| \vec{f'} \right\| = \sqrt{3}$, and the similarity between these two functions is $fsim(f, f') = 2/3$.

With the preceding protein and protein function similarities, the score of the unannotated protein p being annotated by a function $f \in FN(p)$, that is, the contribution of function f to the final prediction results, is defined as

$$score(p, f) = \sum_{p' \in N(p)} \left[sim(p, p') \times \left(\sum_{f' \in F(p')} fsim(f, f') \times \log \frac{N}{n_{f'}} \right) \right] \tag{4.7}$$

where N is the number of all proteins in the whole data set and $n_{f'}$ is the number of proteins in the whole data set that have the function f'. It can be seen from Eq. (4.7) that the value of $fsim(f, f')$ refers to the local impact of available functions within the local domain $N(p)$ on the prediction results, while the value $\log \frac{N}{n_{f'}}$ reflects the global impact of available functions on the prediction results. Intuitively, if a function f' is common to almost all proteins, that is, almost all proteins in the data set have the function f', then its importance and influence decrease, otherwise increase as it is more special.

The iterative function prediction is conducted based on Eq. (4.7). In fact, for each available function $f \in FN(p)$, its contribution to the final prediction results is calculated by the score defined in Eq. (4.7). Therefore all the functions in $FN(p)$ can be ordered by their scores in a descending order, and then the first k functions with the highest scores are selected as the predicted functions of the unannotated protein p. The value of k is determined empirically or by the prediction requirements. In this method k is selected as the average number of functions each protein has in the data set. With the predicted functions of the unannotated protein p, the similarities between the unannotated protein p and its neighbour proteins, that is, $sim(p,p')$ in Eqs. (4.5), (4.7), as well as the function scores are recalculated. With the recalculated scores, all the available functions in $FN(p)$ are reordered and a new prediction is made. This procedure is repeated until the similarities between the unannotated protein p and its neighbour proteins achieve a stable state (Chi and Hou, 2011).

To start the preceding iterative prediction procedure, initial functions must be assigned to the unannotated protein p so that the similarities between the unannotated protein p and its neighbour proteins in Eq. (4.7) can be calculated as per Eq. (4.5). The selection of initial functions for the unannotated protein p is determined by the initial function scores calculated by Eq. (4.7) but with the similarity $sim(p,p')=1$ for any $p' \in N(p)$; that is, for each function $f \in FN(p)$, its initial score is

$$score^{(0)}(p,f) = \sum_{p' \in N(p)} \sum_{f' \in F(p')} \left[fsim(f,f') \times \log \frac{N}{n_{f'}} \right] \tag{4.8}$$

The threshold for initial function selection is set as:

$$\varepsilon = \frac{1}{size(FN(p))} \sum_{f \in FN(p)} score^{(0)}(p,f) \tag{4.9}$$

where $size(FN(p))$ is the number of functions in the set $FN(p)$. The functions with the scores calculated by Eq. (4.8) over the threshold (4.9) are selected as the initial predicted functions of the unannotated protein p.

It is observed from the preceding iterative prediction algorithm that the similarity definition of two proteins $sim(p,p')$ is the key to conducting the function prediction iteratively. If the protein similarity is defined any way other than from protein functions, the prediction algorithm based on Eq. (4.7) is just a normal weighted kNN algorithm, and the prediction cannot be conducted iteratively, and in turn the prediction does not reflect the dynamic features of protein interaction because it is just a one-off process. In other words, this iterative prediction algorithm reflects the dynamic features of protein interaction when predicting functions of unannotated proteins.

4.3 GLOBAL ITERATIVE FUNCTION PREDICTION METHOD

It is observed that the majority of the existing function prediction approaches, including the two iterative prediction methods presented in the preceding sections, exclude the unannotated proteins from the prediction domain. The prediction domain is an interaction subnetwork that consists of annotated proteins required to serve as the information sources for function prediction or to participate in the prediction processes. This exclusion makes the unannotated proteins, as well as their corresponding interaction information in the interaction network, not participate in the prediction process. In other words, a lot of interaction information is discarded when predicting functions from PPIs, and as a consequence the prediction quality could be affected. This consequence might be significant if the PPI data set is in good quality. Furthermore, this exclusion also implies that algorithms that utilize available annotation information (e.g., semantic information-based methods or some classification algorithms) for function prediction are incapable of predicting functions of a protein if most of its neighbour proteins are unannotated.

The global iterative function prediction method introduced in this section is given to address the preceding issues by incorporating unannotated proteins into the prediction processes, while still keeping the merits of iterative approaches in the prediction. This approach makes it possible to exploit global interaction information within the whole interaction data set, rather than just a local subnetwork of interactions, to predict functions of unannotated proteins in the data set. The algorithm starts with identifying all unannotated proteins in the prediction domain (e.g., the whole interaction data set). Then for each unannotated protein, its functions are initialized by an existing prediction algorithm, such as the neighbour counting (NC) algorithm (Schwikowski et al., 2000) or function similarity weighted (FSW) algorithm (Chua et al., 2006). After every unannotated protein in the prediction domain is initially annotated/predicted, the predicted functions in the prediction domain are then iteratively updated until every unannotated protein gets its predicted functions that do not change anymore. In fact, this algorithm consists of four steps: unannotated protein identification, protein function initialization, iterative function updating, and final prediction result selection. The details of these steps are as follows.

4.3.1 Unannotated Protein Identification

In this step, for a target protein P_x for which functions are to be predicted, the unannotated proteins in its prediction domain (e.g., with the NC prediction algorithm, it's the level-1 neighbour proteins of P_x) are first identified. For each of these identified unannotated proteins, if its prediction domain also contains other unannotated proteins, these unannotated proteins are also identified, and the prediction domain of the target protein P_x is extended by incorporating the prediction domains of these unannotated proteins. This identification process continues until all the prediction domains of the identified unannotated proteins do not contain any unannotated proteins. The target protein P_x and all identified unannotated proteins are stored in a set ψ as the new target proteins with a new and bigger prediction domain.

4.3.2 Protein Function Initialization

After unannotated protein identification, this step is to assign initial functions to the identified unannotated proteins. Suppose there are a total of n unannotated proteins (including the main prediction target protein, P_x) in set ψ. When initializing the functions of the i-th protein $P_i (1 \leq i \leq n)$ in ψ, other unannotated proteins in the set $\varphi(\varphi = \psi - \{P_i\})$ and the interactions associated with them are removed from the PPI network. Then functions of the unannotated target protein P_i are initially predicted using an existing protein function prediction algorithm, such as the NC or FSW algorithm. This function initialization process applies to all n unannotated proteins in ψ. For an unannotated target protein which has no annotated neighbours even after the function initialization for other unannotated proteins is finished, it is an isolated protein in the PPI network and will be removed from the prediction domain. The function initialization starts from a randomly selected protein in ψ and in a random order. This initialization is regarded as the first round of prediction, that is, the first iteration.

4.3.3 Iterative Function Updating

After function initialization, the target proteins with initialized nonnull functions are then integrated back into the prediction domain as the annotated proteins and participate in the iterative prediction process. The functions of these proteins will be re-predicted and updated since the protein function distribution in the prediction domain will be changed with the iterative prediction operations.

Details of this step are as follows. Denote N as the total number of iterations before the prediction reaches a stable status.

(1) At the t-th iteration ($t \in (1, N], t = 1$ is for the protein function initialization), for each randomly selected protein $P_i \in \psi(1 \le i \le n)$, a protein function prediction algorithm is applied to re-predict its functions until functions of all proteins in ψ have been re-predicted and updated. Then the current t-th round of prediction ends, and the $(t+1)$-th round of prediction starts.
(2) If re-predicted functions of $P_i \in \psi$ after the $(t+1)$-th round of prediction are different from predicted functions after t-th round of prediction, then its functions are updated as the new re-predicted functions; otherwise keep the functions of P_i unchanged.
(3) Repeat Step 2 until functions of each protein in ψ reach a steady status (i.e., functions of every protein in ψ do not change any more or repeat in a pattern) or after the N-th iteration. Then the iterative function updating ends. If the predicted functions of P_i repeat in a certain pattern after a certain round of predictions and have a repeat span of T (i.e., prediction results from the $(t+T)$-th prediction are the same as the predicted functions from the t-th prediction), then one group of functions from these T groups of predicted functions is randomly selected as the prediction result of P_i.

In each round of iteration, the target proteins for which the function prediction is conducted are selected in a random order. The robustness of the iterative function updating towards different orders of protein selection was experimentally demonstrated in Wang and Hou (2015).

4.3.4 Final Prediction Result Selection

After each round of iteration, the predicted functions of each target protein are recorded. When the iteration ends, for each protein, its final predicted functions are selected statistically from all its predicted functions generated from all rounds of iteration.

Let F be the set of all functions in the PPI network, denoted as $F = \{f^1, f^2, f^3, \ldots, f^K\}$ where $f^i (i = 1, \ldots, K)$ are the functions in the PPI network, and K is the number of functions. For an unannotated protein P_x, its predicted functions after the t-th round of iteration are recorded as a vector $F_{t\,P_x} = \left[f^1_{x,t}, f^2_{x,t}, \ldots, f^K_{x,t}\right]^T$, $f^j_{x,t} = 1\ (j = 1, \ldots, K)$ if the predicted functions of protein P_x from the t-th iteration contain function f^j ($f^j \in F$), otherwise $f^j_{x,t} = 0$. The vector $F_{1\,P_x}$ records the initial functions of P_x. When the iteration reaches the stable status after the M-th iteration, a matrix AF_{P_x} is formed that records all predicted functions of P_x generated from all rounds of the iteration:

$$AF_{P_x} = [F_{1\,P_x}, F_{2\,P_x}, F_{3\,P_x}, \ldots, F_{M\,P_x}]$$

Define a new matrix C_{P_x}:

$$C_{P_x} = AF_{P_x} \times I$$

$$I = [1, 1, \ldots, 1]^T, \quad C_{P_x} = \left[C_{P_x}^{f1}, C_{P_x}^{f2}, C_{P_x}^{f3}, \ldots, C_{P_x}^{fK}\right]^T \quad \text{and} \quad C_{P_x}^{fj} = \sum_{t=1}^{M} f_{x,t}^{j}$$

$(1 \le j \le K)$. The final predicted functions of the unannotated protein P_x are selected as follows:

$$\text{Predicted functions of } P_x = \left\{ f^j \mid f^j \in F,\ C_{P_x}^{fj} \ge \alpha^* \max_{(1 \le j \le K)} \left(C_{P_x}^{fj}\right) \right\} \alpha \in [0, 1]$$

The final predicted functions are actually selected based on the frequency of their occurrences in the whole iteration process, as well as the expected level of occurrence frequency α. The value of α was set to $\alpha = 0.5$ in Wang and Hou (2015).

4.4 DISCUSSIONS

Local and semi-local iterative prediction methods define two new function similarities, as well as two new semantic protein similarities accordingly, based on two dominating protein annotation schemes FunCat and GO, respectively. Actually an iterative prediction method that uses the GO-based similarities can also be adapted to use FunCat-based similarities, and vice versa, because these similarities are defined from the functional relationship/semantic information of proteins, rather than the topological information of a PPI network. Proper definitions of semantic similarity for functions and proteins are the key to conducting iterative predictions.

Local and semi-local iterative prediction methods could be used for different cases where the quality of a protein interaction data set is different. For an interaction data set with good quality data, semi-local prediction method is preferable as it takes into account the local and global influence of a function when iteratively voting for the predicted functions. However, if the quality of an interaction data set is low, the local iterative prediction method is preferable as it takes only the local influence of a function into account when iteratively voting for the predicted functions, so the impact of noise data on the prediction results could be reduced.

The local iterative prediction method can also be extended to include the global influence of functions into the iterative prediction processes, as long as a new function score is defined like Eq. (4.7). Similarly, the semi-global prediction method could also be extended to include the level-2 neighbour proteins into the prediction domain. It is obvious that each extension could incur some costs in terms of computational efficiency and possible influence of noise information on the prediction results. So the possible extension of these two methods also indicates some possible topics for further research.

In the semi-local iterative prediction method, the protein similarity is simply defined based on the indicator function, which can significantly reduce the computational cost in the prediction. In fact, similar to the definition in Eq. (4.3), the similarity between two proteins can also be defined from the similarities of their own functions. The impact of this new similarity on the prediction results is another topic that needs further investigation.

It is obvious that the function similarity in the semi-local iterative method could also be defined in other ways, for example, by using weighed vector elements instead of 1 or 0 to calculate their cosine similarity or using other similarities between vectors to define the function similarities. This chapter mainly focuses on the iterative algorithm and the provision of a framework for further improvement of prediction performance, based on which further research on these topics could be conducted.

The identification of unannotated proteins in the first step of the global iterative prediction method actually depends on the prediction requirements. If only the functions of a specific unannotated protein p are to be predicted, the unannotated proteins in its prediction domain (e.g., its level-1 neighbour proteins) are identified. For each of these identified unannotated proteins, if its prediction domain also contains other unannotated proteins, these unannotated proteins are also identified, and the prediction domain of the protein p is extended to cover the prediction domains of these unannotated proteins. This identification process continues until all the prediction domains of the unannotated proteins do not contain any other unannotated proteins. If there are several specific unannotated proteins whose functions are to be predicted, the unannotated protein identification process is the same for each specific unannotated protein, and the final prediction domain is the merge of the prediction domains of these specific unannotated proteins. The prediction domain could be the whole data set if the functions of all unannotated proteins in the data set are to be predicted.

The prediction algorithm used in the function initialization step of the global iterative prediction method is not necessarily the same as the prediction algorithm used in the iterative prediction process. This makes it possible to exploit the merits of different prediction algorithms to improve the prediction quality and/or efficiency at a reasonable cost. For example, usually a simple or nonsemantic prediction method is used in the function initialization step, while a semantic prediction method is used in the iterative prediction process. This could improve the efficiency of the algorithm in the function initialization step, and in the meantime keep the merits of the semantic iterative prediction algorithm. Although the iterative prediction process could rectify some improper prediction results generated from the initialization, the impact of function initialization on the quality of the final prediction results and to what extent the iterative prediction process can rectify the initialization problems are still unclear at the moment. Further research is worthwhile with respect to these issues.

The global iterative prediction method does not rely on any specific prediction algorithms and similarities. In other words, this method provides an open framework to exploit global interaction information of a protein interaction data set and existing prediction algorithms to more effectively predict functions. The work in Wang and Hou (2015) used three commonly used existing prediction algorithms: NC algorithm (Schwikowski et al., 2000), FSW algorithm (Chua et al., 2006) and function flow algorithm (Nabieva et al., 2005) to evaluate the effectiveness of this method. The evaluation results showed that the algorithms harnessed to the global iterative prediction method outperformed the original algorithms.

REFERENCES

Chi, X., Hou, J., 2011. An iterative approach of protein function prediction. BMC Bioinf. 12, 437.

Chua, H.N., Sung, W.K., Wong, L., 2006. Exploiting indirect neighbours and topological weight to predict protein function from protein-protein interactions. Bioinformatics 22 (13), 1623–1630.

GO Consortium, 2016. Ontology structure. http://geneontology.org/page/ontology-structure (accessed 11.03.16).

Hou, J., Zhu, W., Chen, Y.-P.P., 2013. Dynamically predicting protein functions from semantic associations of proteins. Netw. Model. Anal. Health Inform. Bioinform. 2, 175–183.

Misteli, T., 2001. Protein dynamics: implications for nuclear architecture and gene expression. Science 291, 843–847.

Nabieva, E., Jim, K., Agarwal, A., Chazelle, B., Singh, M., 2005. Whole-proteome prediction of protein function via graph-theoretic analysis of interaction maps. Bioinformatics 21 (Suppl 1), i302–i310.

Ruepp, A., Zollner, A., Maier, D., Albermann, K., Hani, J., Mokrejs, M., Tetko, I., Güldener, U., Mannhaupt, G., Münsterkötter, M., 2004. The FunCat, a functional annotation scheme for systematic classification of proteins from whole genomes. Nucleic Acids Res. 32 (18), 5539–5545.
Schwikowski, B., Uetz, P., Fields, S., 2000. A network of protein-protein interactions in yeast. Nat. Biotechnol. 18 (12), 1257–1261.
Wang, D., Hou, J., 2015. Explore the hidden treasure in protein–protein interaction networks—an iterative model for predicting protein functions. J. Bioinforma. Comput. Biol. 13 (5)1550026 (22 pages).
Zhu, W., Hou, J., Chen, Y.-P.P., 2012. Exploiting multi-layered information to iteratively predict protein functions. Math. Biosci. 236 (2), 108–116.

CHAPTER 5

Functional Aggregation for Protein Function Prediction

5.1 INTRODUCTION

As indicated in (Misteli, 2001), in real biological processes, proteins have high mobility and dynamically interplay to produce a framework which is ever-changing but overall stable. The proteins exchange their biological information and share functions in a dynamic, rather than a static and mono-directed, circumstance. This nature of biological processes implies that protein functions are associated with each other to some extent, although the mechanism of the association is not fully understood yet. For example, some functions in different proteins might be functionally similar or the same in the biological processes. This fact is the basis for function prediction methods that rely on direct interactions between proteins. It is also possible that functions in different proteins are complementary to or reinforce each other to make the biological processes work properly. This fact is the basis for function prediction methods that exploit indirect interactions among proteins to predict protein functions. In other words, above a protein interaction network, logically there is another function association network that is related to the protein interaction network. Both networks are dynamic because of the dynamic features of the biological processes and the critical roles proteins play in the processes. A prediction method could more precisely and reasonably predict functions of unannotated proteins if it can reflect this dynamic feature of protein interactions and function associations in the prediction algorithm. The iterative methods introduced in Chapter 4 address one aspect of this dynamic feature in their prediction algorithms regarding the protein interactions and function associations. Actually, this one aspect focuses only on the mutual and dynamic interactions between the unannotated protein and the available annotated proteins and, in turn, between the predicted function and the available functions of annotated proteins in the prediction domain. The interdependency or dynamic associations among the annotated proteins in the domain are not considered in the prediction processes, which might result in biased prediction results.

New Approaches of Protein Function Prediction from Protein Interaction Networks
http://dx.doi.org/10.1016/B978-0-12-809814-1.00005-4
© 2017 Elsevier Ltd.
All rights reserved.

For a prediction method, only considering the associations between the unannotated protein and the annotated proteins in the prediction domain is based on an implied assumption that the annotated proteins, and in turn their functions, are independent. In fact, if this assumption is true, for an unannotated protein, the prediction results generated from its two annotated neighbour proteins that are not only similar to the unannotated protein but also to each other will be the same as the results generated from its two annotated neighbour proteins that are totally different. That's not a reasonable case to expect. As a matter of fact, from the problem-solving perspective, "one should realize that a set of cases can be complementary in the sense that the experiences represented by the individual cases complement or reinforce each other." "On the other hand, cases can also be redundant in the sense that much of the information is already represented by a smaller subset among them" (Hüllermeier, 2006).

One example in Hüllermeier (2006) illustrates this point. For a yearly rainfall estimation problem at a certain location x_0 (e.g., a city), given the rainfall y_i at locations x_i ($i=1,2,3$) that are the neighours of x_0, what about the rainfall at location x_0 in the two scenarios (a) and (b) shown in Fig. 5.1? It is obvious that for these two scenarios, the y_i should not be combined in the same way for estimating the rainfall at x_0, although the individual instances between x_0 and the x_i are the same. This is because the arrangements or distances of the neighbours are different for these two scenarios (Zhang et al., 1997). Simply predicting the mathematical average $(y_1+y_2+y_3)/3$ as the rainfall at location x_0 seems to be reasonable for the scenario (a), while the same prediction appears to be questionable for the scenario (b). In fact, in scenario (b) the locations x_1 and x_2 are closely neighboured, and the rainfall information at these locations should be partially redundant. Consequently, the weight of joint information from the observations (x_1, y_1) and (x_2, y_2) should not be twice as high as the weight of the information that comes from (x_3, y_3).

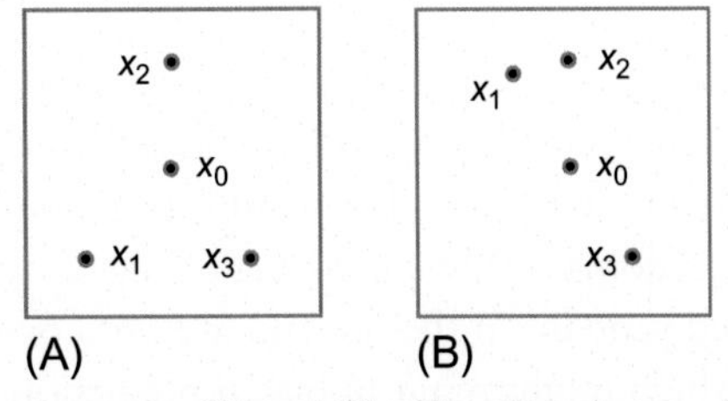

Fig. 5.1 Different arrangements (A) and (B) of locations in two-dimensional space.

In this example, if the location x_0 is considered as an unannotated protein and the other location x_i ($i=1,2,3$) as the annotated proteins, and the distance between the locations as the similarity between proteins, then the rainfall prediction problem could be logically regarded as a function prediction problem, for which the associations between the available annotated proteins (i.e., x_i in this example) should be taken into account.

In fact, applying the point and the discovery in Hüllermeier (2006) to the protein function prediction, it is necessary that the mutual dependencies among neighbour proteins should be taken into account when predicting functions of an unannotated protein. To be more precise, if two or more neighbour proteins are very similar in term of their functions, their contribution to the function prediction of the unannotated protein should not significantly accumulate because they carry very similar functions, while the contribution of those proteins that are not similar should play a more important role in function prediction. In other words, a function prediction algorithm should take into account the contributions of each individual neighbour protein, as well as the correlations among neighbour proteins. To this end, this chapter introduces a new method that uses the Choquet–Integral from fuzzy theory (Hüllermeier, 2006) to aggregate functional correlations among neighbour proteins when predicting functions of unannotated proteins.

5.2 FUNCTIONAL AGGREGATION-BASED PREDICTION

Before providing details of the prediction method, some concepts used in this chapter are defined. For an unannotated protein p_0 whose functions are to be predicted, the set of all its neighbour proteins is denoted as $N(p_0)$. In this method, proteins that have direct interactions with p_0 are the neighbour proteins of p_0. Suppose the number of proteins in $N(p_0)$ is n, that is $N(p_0)=\{p_1, \ldots, p_n\}$. For each protein $p_i \in N(p_0)$ ($i=1, \ldots, n$), the set of its functions is denoted as $F(p_i)$, and the set of all functions from the neighbour proteins is defined as $F=\cup_{i=1}^{n} F(p_i)$. The indicator function $\delta_f(p_i)$ is defined as

$$\delta_f(p_i)=\begin{cases} 1 & \text{if } f \in F(p_i) \\ 0 & \text{otherwise} \end{cases}$$

where $i=1,\ldots,n$, and $p_i \in N(p_0)$.

The basic idea of this prediction method is that for each available function in the function set F, its estimated score of being a predicted function is calculated, and the functions with a higher score are selected as the final predicted functions of the unannotated protein. The calculation of the estimated score for each function will aggregate the correlations among the neighbour proteins via the Choquet–Integral in fuzzy theory. In detail, for a given function $f \in F$, only those neighbour proteins that have the function f are selected to calculate the estimated score of f. For simplicity, denote this protein set as $\{p_1^f, \ldots, p_m^f\}$ where $m \leq n$. Let function $\nu(A)$ be a fuzzy measure on a protein set A. The fuzzy measure $\nu(A)$ measures the degree of closeness among the proteins in set A. The definition of $\nu(A)$ is to be given later. Here, suppose that we have already defined the fuzzy measure function $\nu(A)$. The estimated score of function f being a predicted function of protein p_0 is calculated as

$$p_0^{est(f)} = \sum_{i=1}^{m} \delta_f(p_i)\left(v\left(A_i^f\right) - v\left(A_{i-1}^f\right)\right)$$

where $f \in F$, the set A_i^f is defined as $A_i^f = \left\{p_1^f \ldots p_i^f\right\}$ $(i \leq m)$ which is a part of protein sequence $\{p_1^f \ldots p_m^f\}$ and $A_0^f = \emptyset$. Since $\delta_f(p_i)$ always equals 1 and $v\left(A_0^f\right) = v(\emptyset) = 0$ according to the definition of fuzzy measure (Hüllermeier, 2006), the preceding formula can be simplified as

$$p_0^{est(f)} = v\left(A_m^f\right) \tag{5.1}$$

The following is dedicated to the definition of fuzzy measure $v(A)$ on a protein set A, on which the formula (5.1) is based.

The fuzzy measure $v(A)$ actually measures the protein weight of set A (Hüllermeier, 2006). The measurement is to reflect the fact that evidence coming from a protein set A (i.e., the weight of A) should be discounted if proteins in A are similar among themselves, while the weight of A should be increased if proteins in A are 'diverse' in a certain sense. Therefore the definition of fuzzy measure $v(A)$ depends on the similarity definition between proteins. Here it is assumed that the similarity $sim(p_i, p_j)$ between two proteins p_i and p_j has already been defined (the actual definition of protein similarity will be given later).

Firstly, the diversity $div(A)$ of a protein set A is defined by the sum of pair-wise dissimilarities of proteins in set A:

$$div(A) = \sum_{p_i \neq p_j \in A} \left[1 - sim\left(p_i, p_j\right)\right] \tag{5.2}$$

$div(A) = 0$ if A contains only one protein or A is empty.

Secondly, the similarity contribution of set A, which is made by the individual proteins in A, to the protein p_0 is defined as follows:

$$\mu(A) = \sum_{p_i \in A} sim(p_0, p_i) \tag{5.3}$$

Thirdly, in order to combine the similarity contribution and the protein weight of a protein set A to measure the overall impact of set A on the protein p_0, the fuzzy measure which is not normalized is defined as follows:

$$\bar{v}(A) = \mu(A) + \alpha * div(A) \tag{5.4}$$

where the parameter $\alpha \geq 0$ controls the extent to which correlations among proteins in A are taken into consideration. When $\alpha = 0$, the impact of correlations among proteins in A is completely ignored. The value of α could be determined empirically. In Hou and Chi (2012) the value of α was set as $\alpha = 1.8$ which achieved best prediction results in most cases.

To guarantee the boundary condition of the fuzzy measure (5.4), the final fuzzy measure $v(A)$ is defined as follows:

$$v(A) = \bar{v}(A) / \bar{v}(N(p_0)) \tag{5.5}$$

The following example shows the effect of incorporating the functional aggregation, or the fuzzy measure (5.5), into function prediction. As illustrated in Fig. 5.2, this example predicts the functions of protein p_0 that interacts with three annotated proteins p_1, p_2 and p_3. For demonstration only, suppose that p_1 is annotated by two functions (F_1, F_2), p_2 is annotated by two functions (F_1, F_3), p_3 is annotated by two functions (F_2, F_3), and the similarities between p_0 and its neighbour proteins are 1. Meanwhile, assume $sim(p_1, p_2) = 0.9$, $sim(p_1, p_3) = 0$, and $sim(p_2, p_3) = 0.1$ as shown in the table in Fig. 5.2.

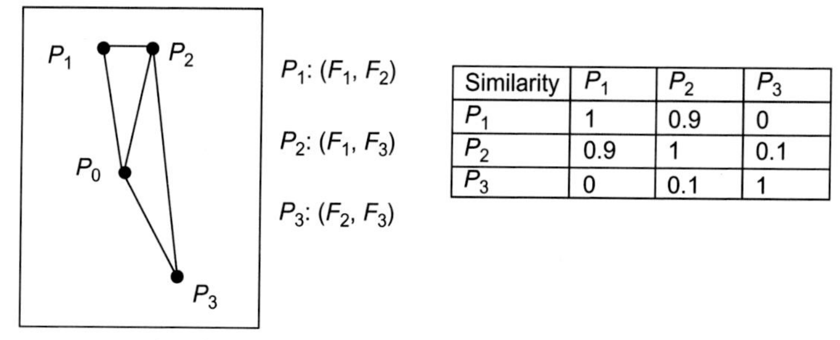

Similarity	P_1	P_2	P_3
P_1	1	0.9	0
P_2	0.9	1	0.1
P_3	0	0.1	1

Fig. 5.2 An example of showing the effect of functional aggregation.

For this example, if the majority voting method (Schwikowski et al., 2000) is used to predict the functions of protein p_0, the scores of all functions F_1, F_2, F_3 are the same as the frequency of each function (i.e., the frequency is 2). In other words, if the correlations among neighbour proteins p_1, p_2, and p_3 in terms of similarity were not considered, it would be impossible to predict which function should be assigned to protein p_0. However, if the functional aggregation is used in the function prediction, that is, Eq. (5.1), the estimated scores of the functions will be different. These scores are listed in the following table ($\alpha = 0.5$ in Eq. (5.4) for this example).

Function	F_1	F_2	F_3
$p_0^{est(f)}$	0.525	0.75	0.725

It can be seen from this table that function F_2 has the highest score, because it comes from proteins p_1 and p_3, which are totally different from each other (i.e., the similarity is 0). By contrast, function F_1 has the lowest score, because it comes from proteins p_1 and p_2, which are quite similar (i.e., the similarity is 90%), which means the information about F_1 is likely to be partially repeated. This example also shows the necessity of incorporating the functional aggregation of proteins into function prediction.

Now it is the time to define a similarity between two proteins, based on which Eqs. (5.2), (5.3) can be calculated for the fuzzy measure of a protein set A, that is, Eq. (5.4) or (5.5). Different from other existing similarities, a new protein similarity is defined semantically from protein function similarities. Suppose a protein p_i is annotated by the functions $\{f_1^{(i)}, \ldots, f_m^{(i)}\}$ and another protein p_j is annotated by the functions $\{f_1^{(j)}, \ldots, f_n^{(j)}\}$. Denote the similarity between two functions $f_x^{(i)}$ and $f_y^{(j)}$ as $FS_{x,y}^{(i,j)}$, and the set of all function similarities of proteins p_i and p_j as $FS^{(i,j)} = \left\{FS_{x,y}^{(i,j)} | x = 1, \ldots, m; y = 1, \ldots, n\right\}$. The similarity $sim(p_i, p_j)$ between two proteins p_i and p_j is defined as the average of their function similarities, that is,

$$sim\left(p_i, p_j\right) = \frac{1}{m \times n} \sum_{x=1}^{m} \sum_{y=1}^{n} FS_{x,y}^{(i,j)} \tag{5.6}$$

The similarity between two protein functions in this method is defined within the Gene Ontology (GO) functional annotation scheme (GO Consortium, 2016). The semantic similarity in a taxonomy (Lin, 1998) is adapted to define the similarity between two functions. Actually, with GO each function is denoted by a GO term, and the similarity between two protein functions is measured by the ratio between the amount of information needed to state the commonality of both functions and the amount of information needed to fully describe two functions. Specifically, suppose functions $f_x^{(i)}$ and $f_y^{(j)}$ are denoted by the GO terms g_x and g_y, respectively; then the similarity $FS_{x,y}^{(i,j)}$ between two functions $f_x^{(i)}$ and $f_y^{(j)}$ is defined as

$$FS_{x,y}^{(i,j)} = \frac{2 \times \log P_{ms}\left(g_x, g_y\right)}{\log P(g_x) + \log P\left(g_y\right)} \tag{5.7}$$

where $P(g_x)$ and $P(g_y)$ are the probabilities that a randomly selected function is annotated by the GO terms g_x and g_y, respectively, and $P_{ms}(g_x, g_y)$ is the probability that a randomly selected function is annotated by the most specific parent GO term of both terms g_x and g_y. More details and calculation examples of these probabilities can be found in Lin (1998).

The protein similarity defined by function similarities, that is, Eqs. (5.6), (5.7), avoids the issues of mismatching of the protein and function similarities in fuzzy measure calculation. Actually, similar to the argument in Hüllermeier (2006), the similarity between two proteins (if it was not defined from their function similarities) is not sufficient to judge whether the functional information in these two proteins is partially repeated. Only in the instance where the proteins, as well as their functions, are similar can we say the functional information in these two proteins is partially repeated. It can be seen from the protein similarity definition (5.6) that if two proteins are similar, their functions are also similar, and vice versa. So these two similarities are unified in one definition, and the similarity mismatching issue is resolved. Hüllermeier (2006) noticed this issue in the fuzzy measure calculation for general cases, but did not provide a solution.

Up to this point, the only problem to be solved in this prediction method is the similarity calculation between the unannotated protein p_0 and its neighbour proteins in Eq. (5.3). This is because the protein similarity is defined from protein function similarities but the functions of p_0 are unknown (actually they are to be predicted). To solve this problem, the iterative prediction approach in Chapter 4 or Chi and Hou (2011) can be adapted to combine the iterative function prediction with the functional aggregation. The details are as follows. Initially assume that the similarities between the unannotated protein p_0 and its neighbour proteins are 1 (i.e., 100%). For each candidate function in the neighbour function set F, use Eq. (5.1) to calculate its estimated score of being a predicted function. The candidate functions with a higher estimated score are selected as the first-round predicted functions of p_0. With the first-round predicted functions, use Eq. (5.1) again to recalculate the estimated scores of all candidate functions and select the second-round predicted functions. This process is repeated until the predicted functions do not change anymore with the iterative prediction process. For each round of prediction, the number of predicted functions is set to be the average number of functions each protein has in the neighbour. As indicated and computationally evaluated on real PPI data sets in Chi and Hou (2011), this iterative prediction operation with functional aggregation of proteins is convergent and able to provide better prediction results compared to existing methods.

5.3 DISCUSSIONS

Even though many methods achieved good function prediction results in some cases, it still remains a challenge to exploit rich information conveyed by PPI networks to improve prediction performance for more general cases in terms of both methodologies and their evaluation (Sharan et al., 2007). The method introduced in this chapter improves function prediction by incorporating correlations among neighbour proteins into protein function prediction via the Choquet–Integral in fuzzy theory. This new prediction method decreases the impact of partially or totally repeated functional information on the final prediction results and, in turn, increases the effectiveness of the prediction method. To make this functional aggregation approach work in function prediction, it is necessary to define a new semantic protein similarity that is based on protein function similarities so that the protein similarities and function similarities are united in the aggregation process.

This functional aggregation approach of function prediction is the first step of incorporating protein correlations into protein function prediction, because the method selects only the first-level (i.e., direct) interaction proteins of an unannotated protein as the neighbour proteins (i.e., the prediction domain). Extending the method to include second-level interaction proteins in the prediction domain, and accordingly developing a new protein similarity and a new prediction algorithm with functional aggregation in a new prediction domain, is one of the ways to improve this method.

The method introduced in this chapter mainly focuses on incorporating the correlations of annotated proteins into prediction. Regarding the prediction process, it adapts only the iterative prediction algorithm for a local prediction domain. Borrowing the ideas from the semi-local and global iterative prediction algorithms in Chapter 4, improving the prediction performance or developing new prediction algorithms by incorporating the global impact information of a function into the final function score calculation in Eq. (5.1) is promising. The global impact of a function could be calculated by using the existing techniques in the information retrieval or other areas. For instance, for a function i, its global impact within the PPI data set could be calculated by $\log(N/n_i)$ where N is the total number of proteins in the PPI data set and n_i is the number of proteins in the PPI data set that have function i. Accordingly, formula (5.1) could be redefined by incorporating the global impact information of available functions.

It is obvious from the method description that once the prediction domain is selected—for example, the first-level or the first- and second-level annotated neighbours of the unannotated protein are selected as the prediction domain—the contribution of the correlations among neighbour proteins to the final prediction results is also fixed; that is, the value of $div(A)$ in Eq. (5.2) is fixed. Actually, if the functional aggregation method in this chapter is combined with the global iterative prediction method in Chapter 4, the value of $div(A)$ in Eq. (5.2) will be dynamic rather than static/fixed in the first two rounds of iteration. This could also be a new algorithm to develop to exploit both the dynamic features and the correlations of protein interactions for more effective function prediction.

Another issue the functional aggregation approach has to face is that a PPI data set usually contains a lot of noise interaction information; that is, some protein interactions are not real. Therefore incorporating the correlations between proteins into prediction might also bring noise information into prediction and, as a consequence, distort the final prediction results. How to incorporate the reliable functional aggregation into prediction or how to combine reliability assessment of protein interactions with prediction is a challenge as well as a starting point from which more effective algorithms could be developed. In this chapter, the functional aggregation via Choquet–Integral is used with only voting-based prediction methods. In fact, measuring correlations among proteins via Choquet–Integral is independent of any specific function prediction methods. Applying Choquet–Integral to other prediction approaches to incorporate correlations among proteins into prediction is another direction toward improving the existing prediction methods.

REFERENCES

Chi, X., Hou, J., 2011. An iterative approach of protein function prediction. BMC Bioinform. 12, 437.

GO Consortium, 2016. Ontology structure. http://geneontology.org/page/ontology-structure (accessed 11.03.2016).

Hou, J., Chi, X., 2012. Predicting protein functions from PPI networks using functional aggregation. Math. Biosci. 240, 63–69.
Hüllermeier, E., 2006. The Choquet-integral as an aggregation operator in case-based learning. Computat. Intell. Theory Appl. 2006, 615.
Lin, D., 1998. An information-theoretic definition of similarity. In: Proceedings of the Fifteenth International Conference on Machine Learning, Madison, WI, pp. 296–304.
Misteli, T., 2001. Protein dynamics: implications for nuclear architecture and gene expression. Science 291, 843–847.
Schwikowski, B., Uetz, P., Fields, S., 2000. A network of protein-protein interactions in yeast. Nat. Biotechnol. 18 (12), 1257–1261.
Sharan, R., Ulitsky, I., Shamir, R., 2007. Network-based prediction of protein function. Mol. Syst. Biol. 3, 88.
Zhang, J., Yim, Y., Yang, J., 1997. Intelligent selection of instances for prediction in lazy learning algorithms. Artif. Intell. Rev. 11, 175–191.

CHAPTER 6

Searching for Domains for Protein Function Prediction

6.1 INTRODUCTION

Noise information in protein interaction data sets is one of the major issues computational prediction methods have to face and deal with. Generally, there are two ways to deal with this issue. One is to assess the reliability of protein interactions in the data sets and remove the unreliable interactions, as discussed in Chapter 2. Another is to reduce or avoid the impact of noise information on the final prediction results, although the noise information cannot be fully identified and removed. The method introduced in this chapter takes the second way to reduce the impact of noise interaction information on prediction results and therefore improve the prediction quality by dynamically searching for the proper prediction domain.

The majority of the existing function prediction approaches consider the prediction domain, that is, a set of proteins from which the functions are predicted, as being unchangeable during the prediction process. For example, the graph theory-based (Vazquez et al., 2003; Karaoz et al., 2004; Nabieva et al., 2005) and Markov random field-based methods (Deng et al., 2003) consider the whole protein interaction network as the prediction domain, while the majority of voting-based methods (Schwikowski et al., 2000; Hishigaki et al., 2001; Chua et al., 2006) select the neighbour proteins of the targeted unannotated protein as the prediction domain, and these domains do not change in prediction processes. Even for clustering-based methods (Brun et al., 2004; Kirac et al., 2006; Pandey et al., 2009; Zhu et al., 2010), once the clusters are finalized, the prediction is then based on these clusters which no longer change. However, the analysis of high-throughput protein interactions indicates that protein interactions identified by experiments usually contain many false-positive interactions, that is, the protein interactions identified by experiments never take place in cells (Deane et al., 2002). In other words, a significant amount of noise information exists in protein interaction data sets. Therefore, if the prediction domain of a prediction method is selected and fixed based on the assumption

New Approaches of Protein Function Prediction from Protein Interaction Networks
http://dx.doi.org/10.1016/B978-0-12-809814-1.00006-6
© 2017 Elsevier Ltd.
All rights reserved.

that all protein interactions in the domain are reliable, the final prediction results will most likely be distorted or incorrect, and a higher prediction accuracy cannot be achieved. This is because the valuable information for function prediction may be overwhelmed by the noise information in the domain.

In this chapter, a prediction domain is regarded as a set of proteins. Generally, some proteins of a prediction domain are functionally related to the unannotated protein, while others are not. The idea of this new prediction method is to find a suitable domain that contains as many related proteins as possible, from which the functions of an unannotated protein can be predicted. To this end, a probabilistic model is proposed to dynamically select a suitable prediction domain and, in turn, dynamically predict functions. Although this method does not aim at removing unreliable interactions from the protein interaction network, it is able to reduce the impact of noise information on the prediction results, and therefore increase the accuracy of function prediction.

6.2 PREDICTION ALGORITHM

This prediction algorithm mainly focuses on the dynamic selection of a suitable prediction domain. Theoretically the dynamic domain selection in this algorithm is suitable to fixed-domain–based prediction methods and is combined with a prediction method to actually implement the dynamic selection of suitable domain and improve the prediction. So the prediction algorithm and domain selection algorithm are mutually dependent, and the dynamic prediction process relies on the combination of these two algorithms. For better understanding, this new dynamic domain selection algorithm and its correlation with a prediction algorithm in prediction, a representative fixed-domain–based prediction algorithm, *FS-Weighted* algorithm (Chua et al., 2006), is used in this chapter to exemplify the prediction algorithm and show this dynamic prediction process.

The FS-Weighted algorithm predicts the functions of an unannotated protein by selecting functions from the neighbour proteins of the unannotated protein. The selected or predicted functions come from the neighbour proteins that have higher similarities with the unannotated protein. Therefore the base of the prediction is the measurement of protein similarity. Let N_p be the set of neighbour proteins of the protein p, including protein p itself. Here a neighbour protein is the protein that directly interacts with protein p in the protein interaction network. The functional similarity (FS) between two proteins u and v is defined as (Chua et al., 2006).

$$S_{\mathrm{FS}}(u,v) = \frac{2|N_u \cap N_v|}{|N_u - N_v| + 2|N_u \cap N_v|} \times \frac{2|N_u \cap N_v|}{|N_v - N_u| + 2|N_u \cap N_v|}, \tag{6.1}$$

where the notation $|\bullet|$ stands for the number of set elements.

In the case where the interaction between proteins u and v is via an intermediate protein x (transitive interaction) in addition to the direct interaction between u and v, the functional similarity between proteins u and v is then defined as a transitive similarity:

$$S_{\mathrm{TR}}(u,v) = \max\left(S_{\mathrm{FS}}(u,v), \max_{x \in N_u} S_{\mathrm{FS}}(u,x) S_{\mathrm{FS}}(x,v)\right) \tag{6.2}$$

With the preceding protein similarities, for the unannotated protein u, the possibility of a function f being selected as a predicted function of protein u is measured by the score

$$Score(f,u) = \sum_{v \in N_u}\left[S_{\mathrm{TR}}(u,v)\delta(v,f) + \sum_{w \in N_v} S_{\mathrm{TR}}(u,w)\delta(w,f)\right], \tag{6.3}$$

where $\delta(p,f) = 1$ if protein p has function f, otherwise $\delta(p,f) = 0$. With the definition (6.3), the prediction domain of the FS-Weighted prediction algorithm for an unannotated protein is the level-1 and level-2 neighbour proteins of the unannotated protein. Here the level-1 neighbour proteins of a protein are those that directly interact with the protein in the PPI network, while the level-2 neighbour proteins are those that do not interact with the protein but directly interact with the level-1 neighbour proteins of the protein. The function prediction using the FS-Weighted algorithm is conducted as follows: for the unannotated protein u, all functions possessed by the proteins in the prediction domain of u are ranked according to their scores calculated by Eq. (6.3), and those functions with higher scores (or with the scores above a predefined threshold) are selected as the predicted functions of the unannotated protein u.

6.2.1 Suitable Domain Selection Algorithm

The suitable domain selection algorithm is based on a condition where initial functions of the unannotated protein have been predicted (the method for predicting the initial functions will be presented in the following section). Suppose the unannotated protein is u, and the set of level-1 and level-2 neighbour proteins of u is the original prediction domain. For simplicity, we still denote this prediction domain as $\boldsymbol{N_u}$. Other notations used for the algorithm description are as follows:

- $q_0 = \{f_1, f_2 \cdots f_r\}$ is the initial predicted functions of u.

- N is the number of proteins in N_u.
- R is the set of proteins in N_u whose functions are relevant to the real functions of u.
- $\overline{R}$ is the complement of R (i.e., the set of proteins whose functions are irrelevant to the real functions of u).
- V is a subset of the proteins in N_u that is possibly a suitable domain for function prediction.
- V_i is the subset of V whose proteins have function f_i.
- n_i is the number of proteins in N_u that possess the function f_i.

For a protein $p_j \in N_u$, whether it is included in a suitable prediction domain is determined by its functional weight with respect to the unannotated protein u in terms of initial predicted functions, which is defined as

$$PW\left(p_j, q_0\right) = \sum_{i=1}^{r} \delta\left(p_j, f_i\right) \times \left| \left(\log \frac{P(f_i|R)}{1 - P(f_i|R)} + \log \frac{1 - P(f_i|\bar{R})}{P(f_i|\bar{R})} \right) \right|, \tag{6.4}$$

where r is the number of functions in q_0, and

$$P(f_i|R) = \frac{|V_i| + \frac{n_i}{N}}{|V| + 1}, \quad P(f_i|\bar{R}) = \frac{n_i - |V_i| + \frac{n_i}{N}}{N - |V| + 1}.$$

$|V_i|$ and $|V|$ are the number of proteins in the sets V_i and V, respectively. $P(f_i|R)$ stands for the probability of function f_i being presented in a protein selected randomly from the set R. $P(f_i|\bar{R})$ is analogous to $P(f_i|R)$. Recalling that $P(f_i|R) + P(\bar{f}_i|R) = 1$ where $P(\bar{f}_i|R)$ stands for the probability of function f_i not being presented in a protein selected randomly from the set R, it can be seen that Eq. (6.4) measures the suitability of the protein p_j being included in the prediction domain by measuring the ratios between the probabilities of its functions being presented in the suitable protein set and the probabilities of its functions not being presented in the suitable protein set.

6.2.2 Dynamic Function Prediction Algorithm

To select a suitable prediction domain for function prediction, it is necessary to select a possible prediction domain V required by Eq. (6.4). In this algorithm, the selection of a possible prediction domain is based on the functional semantic similarity between the unannotated protein and its neighbour proteins. The functional semantic similarity between two proteins is defined from the similarities of their functions. As discussed in the previous chapters, protein functions are annotated by existing schemes, such as Gene Ontology (GO) (GO Consortium, 2016) or Functional Catalogue (FunCat) (Ruepp et al., 2004). In this algorithm, FunCat is used as an example annotation scheme to define a functional semantic similarity between two proteins. With FunCat, a function is expressed numerically by up to six layers. The digital number at each layer represents a specific function category or function. For example, the function '*Xanthine Catabolism*' is expressed as 01.03.01.01.03 with FunCat scheme. The deeper a function's layer achieves, the more specific the function is. Suppose protein u has functions $\{\boldsymbol{f_1^{(u)}}, \boldsymbol{f_2^{(u)}} \ldots \boldsymbol{f_m^{(u)}}\}$ and another protein v has functions $\{\boldsymbol{f_1^{(v)}}, \boldsymbol{f_2^{(v)}} \ldots \boldsymbol{f_n^{(v)}}\}$, the similarity between two functions $\boldsymbol{f_i^{(u)}}$ and $\boldsymbol{f_j^{(v)}}$ $(\boldsymbol{i}=1,\ldots,\boldsymbol{m}$ and $\boldsymbol{j}=1,\ldots,\boldsymbol{n})$ is defined as

$$FS_{i,j}^{(u,v)} = \frac{\sum_{k=1}^{l} k^2}{\sum_{g=1}^{h} g^2}, \quad l, h \in [1, 6], \tag{6.5}$$

where l is the number of common sequent layers that functions $f_i^{(u)}$ and $f_j^{(v)}$ share from the first layer in terms of FunCat notation, and h is maximum number of layers that two functions $f_i^{(u)}$ and $f_j^{(v)}$ have. Then the functional semantic similarity between two proteins u and v is defined as

$$PS(u, v) = \frac{1}{m \times n} \sum_{i=1}^{m} \sum_{j=1}^{n} FS_{i,j}^{(u,v)}. \tag{6.6}$$

With the preceding algorithms and definitions, the function prediction algorithm with the dynamic prediction domain selection is described in the following steps:

Step 1: Predict initial functions of the targeted unannotated protein u
- Calculate the score of each function f possessed by the neighbour proteins of protein u using Eq. (6.3);
- Rank the functions according to their scores in a descending order and select the first r functions as the initial predicted functions of u (i.e., $q_0 = \{f_1, f_2 \ldots f_r\}$).

Step 2: Select a *possible prediction domain V*
- Calculate the similarity of each neighbour protein with the unannotated protein u, using Eq. (6.6);
- Rank the neighbour proteins according to their similarities in a descending order, and select the first s neighbour proteins as a *possible prediction domain V*.

Step 3: Select a *suitable prediction domain SD*

- Calculate the protein weight of each neighbour protein with respect to the unannotated protein *u*, using Eq. (6.4);
- Rank the neighbour proteins according to their weights in a descending order and select the first *s* neighbour proteins as a *suitable prediction domain SD.*

Step 4: Predict functions from the domain *SD*

- Calculate the score of each function *f* possessed by the proteins in the *SD* using Eq. (6.3);
- Rank the functions according to their scores in a descending order and select the first *r* functions as the new predicted functions of *u* (i.e., $q_0' = \{f_1', f_2', \ldots, f_r'\}$);
- If $q_0' = q_0$, then stop the prediction operation and the current q_0' is the set of final predicted functions of *u*; otherwise, replace q_0 with q_0' and go back to Step 2.

It can be seen from the preceding algorithm steps that the prediction process, as well as the process of selecting a suitable prediction domain, is dynamic until the predicted functions are stable; that is, they no longer change.

This dynamic prediction algorithm is convergent; that is, the predicted functions will be stable after finite rounds of iterative prediction operations. In fact, with the prediction method (i.e., Steps 1 and 4), the functions, which occur more frequently within the neighbour proteins and are possessed by the neighbour proteins that are more similar to the unannotated protein, are most likely to be selected as the predicted functions. The method of a suitable domain selection (i.e., Steps 2 and 3) is to verify whether the predicted functions are most likely located in those proteins that are functionally related to the unannotated protein. Therefore those predicted functions confirmed by one round of prediction will be kept in the next round of prediction, while other functions will be removed or brought into other rounds of prediction for further verification. Since the number of neighbour proteins is finite, after a finite number of prediction rounds, all possible domains will be scanned by the algorithm, and the predicted functions will be stable. The computation experiments on real data sets in (Hou and Jiang, 2013) showed that normally after two to three rounds of dynamic prediction operations, the predicted functions became stable.

The preceding dynamic function prediction algorithm refers to two parameters: *r*, which is the number of predicted functions, and *s*, which is the number of proteins in a suitable prediction domain. What value should be set for the parameter *r* generally depends on the prediction expectation.

For example, if a prediction is expected to provide five predicted functions, *r* is set to five accordingly, or the value of *r* could be set by running the algor ithm on a training data set with different *r* values and choosing a value for *r* that achieves the highest prediction average precision. Generally, the average number of functions each protein has in the data set can be set as a benchmark for choosing the value of *r*. Regarding the value of *s*, usually it is set as one-third (1/3) of the neighbour protein numbers. This setting is based on the research result in (Deane et al., 2002), which indicated that generally only 30–50% of the interactions identified by high-throughput experiments were biologically relevant. The computational experiments in (Hou and Jiang, 2013) also showed that the prediction achieved the highest precision and recall with this *s* value setting.

6.3 DISCUSSIONS

The dynamic function prediction of this algorithm is implemented by the dynamic selection of a suitable prediction domain. It can be seen from the algorithm steps that the dynamic selection of a suitable prediction domain is relatively independent of the prediction operations in terms of similarity measurement and models. In fact, the function prediction operations are performed at Step 1 and Step 4 of the algorithm, which are based on the protein similarity and function score that rely on the topological structure information of the protein interaction network—that is, the Eqs. (6.1), (6.2) and (6.3)—while the dynamic selection of a suitable prediction domain is conducted at Step 2 and Step 3 of the algorithm, which are based on the semantic functional similarity of proteins and the probability model, that is, the Eqs. (6.4), (6.5) and (6.6). This independency makes it flexible to combine the algorithm of dynamic prediction domain selection with other function prediction algorithms to improve the prediction quality. On the other hand, this new dynamic prediction algorithm also provides a framework showing within it how a static function prediction algorithm (e.g., FS-Weighted algorithm) is combined with the dynamic prediction domain selection to form a dynamic prediction algorithm, as well as how the topological structure information of a protein interaction network and the functional semantic information of proteins are integrated into the protein function prediction.

The relative independency of the function prediction and the dynamic prediction domain selection also make it possible to develop new algorithms. For instance, the functional semantic similarity between proteins defined by

Eqs. (6.5), (6.6) is based on the FunCat annotation scheme because of its computational simplicity. Actually, this semantic protein similarity could also be defined from the GO annotation scheme just like the ones in Chapters 4 and 5. The computational experiments in (Hou and Jiang, 2013) showed the effectiveness of this prediction algorithm when using the GO annotation–based semantic similarity between proteins. Similarly, the prediction algorithm could also be other ones, such as the majority voting algorithm in Schwikowski et al. (2000), the iterative algorithms in Chapter 4 or the functional aggregation algorithm in Chapter 5.

This function prediction algorithm does not directly deal with the issues of protein interaction reliability in the prediction. Instead, it tries to avoid or reduce the impact of possible unreliable interactions on final prediction results by dynamically selecting suitable prediction domains. This treatment might not be enough to significantly reduce the impact of noise information on the prediction results. Incorporating the interaction reliability assessment into the prediction processes might be a promising way to improve the effectiveness of the prediction algorithm.

When selecting neighbour proteins to form a new suitable prediction domain in the algorithm, the number of selected proteins *s* is set as one-third (1/3) of the total number of neighbour proteins (original prediction domain). This setting is based on the observations from the existing empirical research with respect to the existing protein interaction databases. It is worth noting that the definition of the functional semantic similarity between proteins, that is, Eq. (6.6), also has an impact on the quality of the selected new prediction domain. Since the quality of the new suitable prediction domain has a great impact on the final prediction quality, how to more reasonably and objectively determine the value of the parameter *s* for a specific protein interaction database or a specific prediction domain and how to define the protein semantic similarity more properly are interesting topics for further research.

REFERENCES

Brun, C., Chevenet, F., Martin, D., Wojcik, J., Guénoche, A., Jacq, B., 2004. Functional classification of proteins for the prediction of cellular function from a protein-protein interaction network. Genome Biol. 5 (1), R6.

Chua, H.N., Sung, W.K., Wong, L., 2006. Exploiting indirect neighbours and topological weight to predict protein function from protein-protein interactions. Bioinformatics 22 (13), 1623–1630.

Deane, C.M., Salwinski, L., Xenarios, I., Eisenberg, D., 2002. Protein interactions: two methods for assessment of the reliability of high throughput observations. Mol. Cell. Proteomics 1 (5), 349–356.

Deng, M., Zhang, K., Mehta, S., Chen, T., Sun, F., 2003. Prediction of protein function using protein-protein interaction data. J. Comput. Biol. 10 (6), 947–960.
GO Consortium, 2016. Ontology structure. http://geneontology.org/page/ontology-structure (accessed 11.03.16).
Hishigaki, H., Nakai, K., Ono, T., Tanigami, A., Takagi, T., 2001. Assessment of prediction accuracy of protein function from protein-protein interaction data. Yeast 18 (6), 523–531.
Hou, J., Jiang, Y., 2013. Dynamically searching for a domain for protein function prediction. J. Bioinforma. Comput. Biol. 11 (4), 1350008.
Karaoz, U., Murali, T.M., Letovsky, S., Zheng, Y., Ding, C., Cantor, C.R., Kasif, S., 2004. Whole-genome annotation by using evidence integration in functional-linkage networks. Proc. Natl. Acad. Sci. U. S. A. 101, 2888–2893.
Kirac, M., Ozsoyoglu, G., Yang, J., 2006. Annotating proteins by mining protein interaction networks. Bioinformatics 22 (14), e260–e270.
Nabieva, E., Jim, K., Agarwal, A., Chazelle, B., Singh, M., 2005. Whole proteome prediction of protein function via graph-theoretic analysis of interaction maps. Bioinformatics 21 (Suppl 1), i302–i310.
Pandey, G., Myers, C.L., Kumar, V., 2009. Incorporating functional inter-relationships into protein function prediction algorithms. BMC Bioinf. 10, 142.
Ruepp, A., Zollner, A., Maier, D., Albermann, K., Hani, J., Mokrejs, M., Tetko, I., Güldener, U., Mannhaupt, G., Münsterkötter, M., 2004. The FunCat, a functional annotation scheme for systematic classification of proteins from whole genomes. Nucleic Acids Res. 32 (18), 5539–5545.
Schwikowski, B., Uetz, P., Fields, S., 2000. A network of protein-protein interactions in yeast. Nat. Biotechnol. 18 (12), 1257–1261.
Vazquez, A., Flammini, A., Maritan, A., Vespignani, A., 2003. Global protein function prediction from protein–protein interaction networks. Nat. Biotechnol. 21, 697–700.
Zhu, W., Hou, J., Chen, Y.-P.P., 2010. Semantic and layered protein function prediction from PPI networks. J. Theor. Biol. 267 (2), 129–136.

CHAPTER 7

Protein Function Prediction from Functional Connectivity

7.1 INTRODUCTION

Protein interaction network provides a framework that brings different kinds of information together, such as topological information of the network, correlations between proteins, and functional relationships between proteins. How to effectively reveal and exploit various information hidden in protein interaction networks to more precisely predict functions of unannotated proteins has been a challenge. The majority of similarity-based prediction methods, such as those introduced in the previous chapters, consider the contribution of annotated proteins to the final prediction results is individual. In other words, whether a function processed by an annotated protein could be a predicted function of a target unannotated protein generally depends on the similarity between the annotated protein and the target unannotated protein. The roles of the correlation between proteins in function prediction are not fully investigated. The functional aggregation method introduced in Chapter 5 for protein function prediction is innovative in trying to exploit protein correlations to improve the prediction performance. However, it takes into account only the correlations between proteins in terms of similarity.

In addition to the similarity, the correlations between proteins can also be interpreted in terms the importance within a protein interaction data set or network. It is well known in the network model that the importance of a node could be measured by its degree, that is, the number of edges incident on the node. In other words, the correlation between a node and other nodes in the network determines its importance within the network. Regarding the protein interaction network, a node is a protein, and an edge is an interaction between two proteins. The degree of a protein in a protein interaction network is also known as the protein connectivity (Maslov and Sneppen, 2002). If an annotated protein is more important—that is, has a higher connectivity—within a protein interaction network, the functions it possesses are more likely to be selected as the predicted functions of

New Approaches of Protein Function Prediction from Protein Interaction Networks
http://dx.doi.org/10.1016/B978-0-12-809814-1.00007-8
© 2017 Elsevier Ltd.
All rights reserved.

unannotated proteins, as these functions are endorsed by other proteins. This kind of correlation information between proteins could be incorporated into the prediction processes to more reasonably and effectively predict functions.

In protein interaction networks, most proteins interact with few partners, whereas a small but significant proportion of proteins, the hubs, interact with many partners. It is widely recognized that hub proteins, that is, proteins with high connectivity, play more important biological roles than non-hub proteins, that is, proteins with low connectivity. In general, protein interaction networks are tolerant toward random protein removal, but they are very sensitive to the targeted removal of hubs. For example, knockouts of yeast genes encoding hubs are approximately threefold more likely to cause lethality than knockouts of non-hubs. This indicates that there exists a connection between genetic robustness and the topology of protein–protein interaction networks (Han et al., 2004).

Han et al. (2004) investigated how the hubs might contribute to robustness and other cellular properties for protein–protein interactions that are dynamically regulated in time and space. They identified two types of hubs: 'party' hubs and 'date' hubs (Fig. 7.1). Party hubs interact with most of their partners at the same time (i.e., party hubs are static), while date hubs bind their different partners at different times or locations (i.e., date hubs are dynamic). The date hubs connect biological processes (or modules) to each

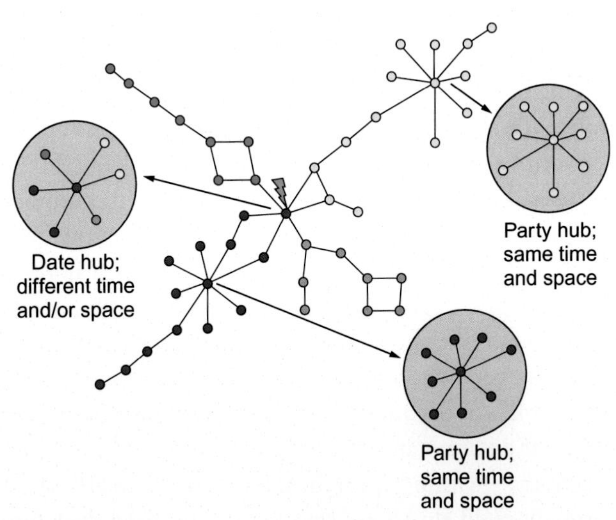

Fig. 7.1 Party and date hub proteins in protein interaction networks (Han et al., 2004).

other and thus organize the proteome, while the party hubs function inside the modules. Hub proteins have proved to be highly conserved and play a pivotal role in the whole protein interaction network (Han et al., 2004). Therefore, in protein function prediction, it is natural and necessary to take the protein connectivity into consideration.

However, in function prediction, simply considering the protein connectivity in a prediction algorithm might not really reflect the functional importance of proteins in the prediction processes. This is because an edge in a protein interaction network is static and only indicates that most likely there is an interaction between two proteins. It neither reveals any information about how tight or strong the interaction is in terms of the functionality between two proteins, nor indicates whether the interaction is functionally reliable. Therefore, instead of directly using protein connectivity in function prediction, the method introduced in this chapter defines and uses a new concept, protein functional connectivity, to measure the functional strength (rather than the topological strength) of a protein's impact on its neighbours and, in turn, determine the functional importance of proteins in a protein interaction network. The details of the function prediction algorithm that is based on this new protein functional connectivity are presented in the following section.

7.2 FUNCTION PREDICTION ALGORITHM

This prediction algorithm is semantic similarity-based; that is, the prediction is based on the semantic similarity between two proteins that is defined from the similarities of their own functions. Furthermore, protein functional connectivity is based on the protein semantic similarities as well. Therefore the definitions of the function similarity and the protein semantic similarity are presented first. For computational simplicity, this algorithm adopts the FunCat (Ruepp et al., 2004) scheme for protein function annotation. Excluding experimentally uncharacterized proteins, FunCat contains six main functional categories: metabolism, information pathways, transport, perception and response to stimuli, developmental processes and localization. A protein function is described in no more than six layers of digital numbers; for example, the FunCat annotation '10.03.01.01.01' denotes the function '*G1 phase of mitotic cell cycle*'. According to the definition of the FunCat scheme, a function described in more layers is more specific than a function described in fewer layers.

Function similarity represents the similarity between two functions. With the FunCat annotation scheme, for any two functions f and f', denote the number of their common shared layers in FunCat as $cl(f, f')$ where $0 \leq cl(f,f') \leq 6$, and the number of their maximum function layers as $ml(f,f')$ where $1 \leq ml(f,f') \leq 6$; then their similarity $fsim(f,f')$ is defined as

$$fsim(f,f') = \frac{\sum_{i=0}^{cl(f,f')} i^2}{\sum_{j=1}^{ml(f,f')} j^2}$$

For example, two functions 'glycogen anabolism' and 'glycogen catabolism' in biological process, which are denoted as '01.05.03.01.04' and '01.05.03.01.07', respectively, with the FunCat scheme have four commonly shared layers, that is, '01.05.03.01', $cl(f, f')=4$ and $ml(f, f')=5$; therefore the similarity between these two functions is 0.55.

Based on the preceding function similarity definition, the semantic similarity between two proteins is defined as the weighted average of their function similarities. Specifically, let $F(p)$ be the function set of protein p, then the semantic functional similarity $FS(p, p')$ between two proteins p and p' is defined as

$$FS(p, p') = \frac{\sum_{f \in F(p)} \sum_{f' \in F(p')} fsim(f,f')}{\sum_{f \in F(p)} \sum_{f' \in F(p')} [2 - cl(f,f')/ml(f,f')]} \tag{7.1}$$

7.2.1 Protein Functional Connectivity

The protein functional connectivity (FC) of a protein defines the strength of a protein's impact on its neighbours, or the endorsement a protein receives from its neighbours. So the functional connectivity also indicates the functional importance of a protein within a protein interaction network. Different from the connectivity concept in the conventional network model, the functional connectivity of a protein considers not only the degree of the protein (node) in the protein interaction network but also the semantic similarity between the protein and its neighbours. Actually, for a protein p, denote the set of its neighbours as $N(p)$, then the functional connectivity $FC(p)$ of p is defined as

$$FC(p) = 1 - \prod_{\substack{p' \in N(p) \\ FS(p,p') \neq 1}} (1 - FS(p, p')) \tag{7.2}$$

Higher FC value of a protein suggests that the protein has closer functional correlations with its neighbours in a protein interaction network.

7.2.2 Function Addition

Function addition (FA) measures the correlation strength between a function and the functions of a protein, or the functional endorsement a function receives from the functions of a protein. For a function f, its function addition with respect to a protein p is defined as

$$FA(f,p) = \sum_{f' \in F(p)} fsim(f,f') \tag{7.3}$$

Generally $FA\ (f,\ p) \geq 0$.

With the preceding FS, FC and FA definitions in Eqs. (7.1), (7.2) and (7.3), for an unannotated protein p, whether a function f from the prediction domain can be selected as a predicted function of p is determined by its functional score with respect to the protein p, which is defined as

$$Score(f,p) = \sum_{p' \in N(p)} FA(f,p') \times FC(p') \times FS(p,p') \tag{7.4}$$

This score definition indicates that whether a function f can be selected as a predicted function of an unannotated protein p depends on three factors: whether the function f is endorsed by the neighbours of p (i.e., $FA(f,p')$), whether the neighbours are functionally important (i.e., $FC(p')$) and whether the neighbours are semantically similar to p (i.e., $FS(p,p')$). The higher the score, the more likely the function f being of a predicted function. Actually, to predict functions for an unannotated protein, all functions from the prediction domain (e.g., the neighbours of the unannotated protein) can be ranked by their functional scores with respect to the unannotated protein in a descending order. Those functions with higher functional scores are then selected as the predicted functions.

7.2.3 Protein Function Prediction

As previously indicated, function prediction is based on the functional score (7.4) of the functions in the prediction domain. In this algorithm, the neighbours of an unannotated protein p, that is, the set $N(p)$ in Eq. (7.4), are defined as the level-1 and level-2 neighbour proteins of p. This definition is based on the observation in (Chua et al., 2006) that a protein is likely to share functions with its level-1 and level-2 neighbours in a protein interaction network. Here level-1 neighbour proteins are those that directly interact with the unannotated protein p, while level-2 neighbour proteins are those that directly interact with the level-1 neighbour proteins of p but do not directly interact with the unannotated protein p.

Using Eq. (7.4) for function ranking and prediction requires the calculation of the semantic similarity $FS(p, p')$ between the unannotated protein p and its neighbour p', which relies on their function similarities. However, the functions of the unannotated protein p are unknown and yet to be predicted. To kick off the prediction, it is necessary to assign initial functions to unannotated proteins in the protein interaction network. Eq. (7.4) then can be used iteratively to predict functions. The function initialization of unannotated proteins in the network could be done by using an existing function prediction algorithm that is based on the topological structure information of the protein interaction network, such as the FS-Weighted algorithm in Chua et al. (2006). The prediction algorithm based on Eq. (7.4) is described in the following steps.

Step 1: Calculate the average number of functions each annotated protein has in the protein interaction data set. Denote this average number as N_f which will be used for selecting predicted functions. Construct the set F of all functions in the protein interaction data set.

Step 2: Initialize functions of all unannotated proteins in the protein interaction data set using an existing prediction algorithm that is based on the topological information of the protein interaction network.

Step 3: For each unannotated protein p in the interaction data set, do the following:

Step 3.1: Calculate $FS(p, p')$ and $FC(p')$ for $p' \in N(p)$ using Eqs. (7.1) and (7.2), respectively.

Step 3.2: For each $f \in F$, calculate $FA(f, p')$ for all $p' \in N(p)$ and $Score(f, p)$ of the function f with respect to p using Eqs. (7.3) and (7.4), respectively.

Step 3.3: Rank all functions in F according to their functional scores in a descending order, and select the first N_f functions as the predicted functions of p.

Step 4: If the predicted functions of each unannotated protein are stable, that is, the same as the predicted functions from the previous round of prediction, then the prediction stops; otherwise, update the functions of all unannotated proteins by the current predicted functions and go to Step 3.

In the algorithm, the first round of prediction is the function initialization for all unannotated proteins in the data set. It can be seen from the algorithm that, because of the unannotated proteins also being included in the neighbours of a target unannotated protein, the FS, FC and FA values in Eq. (7.4) are updated iteratively because the predicted functions of unannotated proteins are iteratively updated with the prediction processes. This ensures that the iterative prediction processes run successfully. The computational evaluations of this algorithm on real protein interaction data sets in (Zhu et al., 2013) showed the convergence and effectiveness of this algorithm.

7.3 DISCUSSIONS

The protein function prediction algorithm introduced in this chapter defines a new concept and metric to measure the functional importance of a protein within a protein interaction network, and incorporates the functional importance of proteins into an iterative function prediction process. The function similarity, as well as the semantic functional similarity between proteins, defined in this algorithm is based on the FunCat annotation scheme for computational simplicity. Actually, the function similarity and the corresponding semantic functional similarity between proteins can also be defined based on other function annotation schemes, such as GO (GO Consortium, 2016), like those defined in Chapters 4 and 5.

This algorithm solves a critical problem regarding the neighbour formation of the target unannotated protein. Usually in conventional prediction methods, unannotated proteins are excluded from the neighbours as their functions are yet to be predicted. This exclusion also removes some valuable protein interaction information from the data set, which in turn affects the prediction quality to some extent. In this new algorithm, unannotated proteins are included in the neighbours of the target unannotated protein, participate in the prediction processes from the beginning and update themselves simultaneously with other proteins (annotated and unannotated) in the data set. This prediction process therefore more reasonably reflects the interaction nature between proteins in the network, and could produce better prediction results.

As indicated in Section 7.2, the neighbours of an unannotated protein are defined as the level-1 and level-2 neighbour proteins of the protein. It is obvious that the neighbour of an unannotated protein can also be extended to include level-3 or even higher level proteins in the neighbour, but as Chua et al. (2006) observed, using higher level neighbours in function prediction also increases the chance of including erroneous interaction information.

The selection condition N_f used in this algorithm to select ranked functions is defined as the average number of functions each annotated protein has in the whole data set. Once this selection condition is determined, it will not change during the prediction. Although it is easy to determine this selection condition, this static number is not flexible and might be unable to reflect the real situation of each unannotated protein because for different unannotated proteins, the number of their real functions might be different. This selection condition actually could be defined locally for each specific unannotated protein; for example, it could be defined as the average number of functions each neighbour protein has with respect to the specific unannotated protein. It could also be defined dynamically based on the functional scores of the candidate functions; for instance, it could be dependent on the score changing rate based on which the algorithm determines where to stop the selection from the list of ranked candidate functions.

In Step 3.2 of the algorithm, all available functions of the data set are the candidates of predicted functions. In fact, the candidate function *f* could be selected locally from the neighbour functions of the unannotated protein. The current algorithm tries to select a candidate predicted function globally from the whole function set of the protein interaction data set. This approach might be able to predict the functions that are not in the neighbours of the unannotated protein. However, the computational cost is probably an issue. For selecting candidate functions locally from the neighbour functions, the computational cost could be reduced, while some real functions of the unannotated protein might not be predicted. This is because usually for a protein in an interaction network, only about 71% of its functions can be found in its level-1 or level-2 neighbours (Chua et al., 2006). So there are still some functions that cannot be predicted from the neighbour functions. Whether to use a global or local way to select candidate functions in prediction depends on the purpose of the algorithm. If the coverage (i.e., to predict as many real functions as possible) is the major concern of the prediction, the global way is a proper choice. If the prediction is mainly concerned about the functions that are closely related to the neighbour proteins, then the local way is suitable.

This algorithm exploits the entire interaction data set to predict functions. Therefore the quality of the data set has a significant impact on the prediction quality. The noise information in the protein interaction data set is still a challenge for this algorithm, as well as other prediction algorithms. Although there exist algorithms for evaluating the reliability of protein interactions such as those introduced in Chapter 2 or other methods (Pena-Castillo et al., 2008; Wachi et al., 2005; Simonis et al., 2004; Kelley and Ideker, 2005; Lu et al., 2005) that try to integrate data sets from different public organizations to form a reliable interaction data set, how to incorporate the interaction reliability weight into the prediction algorithm to improve prediction quality or reduce the impact of noise information on prediction results is still one of the possible research topics that rely on the protein functional connectivity.

REFERENCES

Chua, H.N., Sung, W.K., Wong, L., 2006. Exploiting indirect neighbours and topological weight to predict protein function from protein-protein interactions. Bioinformatics 22 (13), 1623–1630.

GO Consortium, 2016. Ontology structure. http://geneontology.org/page/ontology-structure (accessed 11.03.16).

Han, J.-D.J., Bertin, N., Hao, T., Goldberg, D.S., Berriz, G.F., Zhang, L.V., Dupuy, D., Walhout, A.J.M., Cusick, M.E., Roth, F.P., Marc Vidal, M., 2004. Evidence for dynamically organized modularity in the yeast protein–protein interaction network. Nature 430, 88–93.

Kelley, R., Ideker, T., 2005. Systematic interpretation of genetic interactions using protein networks. Nat. Biotechnol. 23, 561–566.

Lu, L.J., Xia, Y., Paccanaro, A., Yu, H., Gerstein, M., 2005. Assessing the limits of genomic data integration for predicting protein networks. Genome Res. 15 (7), 945–953.

Maslov, S., Sneppen, K., 2002. Specificity and stability in topology of protein networks. Science 296, 910–913.

Pena-Castillo, L., Tasan, M., Myers, C.L., et al., 2008. A critical assessment of Mus musculus gene function prediction using integrated genomic evidence. Genome Biol. 9 (Suppl 1), S2.

Ruepp, A., Zollner, A., Maier, D., Albermann, K., Hani, J., Mokrejs, M., Tetko, I., Güldener, U., Mannhaupt, G., Münsterkötter, M., 2004. The FunCat, a functional annotation scheme for systematic classification of proteins from whole genomes. Nucleic Acids Res. 32 (18), 5539–5545.

Simonis, N., van Helden, J., Cohen, G.N., et al., 2004. Transcriptional regulation of protein complexes in yeast. Genome Biol. 5, R33.

Wachi, S., Yoneda, K., Wu, R., 2005. Interactome–transcriptome analysis reveals the high centrality of genes differentially expressed in lung cancer tissues. Bioinformatics 21, 4205–4208.

Zhu, W., Hou, J., Chen, Y.-P.P., 2013. Semantically predicting protein functions based on protein functional connectivity. Comput. Biol. Chem. 44, 9–14.

CHAPTER 8

Conclusions

8.1 SUMMARY

Proteins, which are large, complex molecules of biological tissues, are the most essential and important molecules of life because they play major structural and functional roles in a cell to make organs and the body work. Correctly annotating protein functions is therefore necessary and greatly helps human beings understand various biological processes, phenomena and even how to develop new methods or products to control and treat diseases.

Although functions of some proteins have been identified and annotated via biological experiments, there are still many proteins whose functions are yet to be annotated because of the limitations of existing methods and the high cost of experiments. On the other hand, new high-throughput technologies have generated a large amount of biological data of many species, such as microarray data, gene sequencing data and protein interaction data. These rapidly expanding biological data volumes make it urgent and possible to use computational methods to analyse and exploit the rich information hidden in the massive data to deeply and systematically explore the secrets of various biological processes and phenomena, especially in predicting the functions of unannotated proteins. However, how to interpret various available biological data and effectively predict protein functions still remains a big challenge in this post-genomic era.

Numerous computational methods have been developed to predict protein functions from various biological data sources. Among them the methods that exploit protein–protein interaction data to predict protein functions have proved to be feasible and effective, because protein–protein interactions more reasonably reflect the nature of protein functions; that is, a protein almost never performs its functions in isolation; instead, it usually interacts with other proteins in order to accomplish a certain function (Misteli, 2001). In addition, protein interactions which can be modelled as a network also provide a flexible platform based on which various information could be brought together for protein function prediction.

This book does not intend to be an encyclopaedia or a survey of computational methods for protein function prediction. Instead, it focuses on the

New Approaches of Protein Function Prediction from Protein Interaction Networks
http://dx.doi.org/10.1016/B978-0-12-809814-1.00008-X
© 2017 Elsevier Ltd.
All rights reserved.

innovative prediction approaches and methods that have not been systematically summarised and discussed. Moreover, at the end of each chapter, this book also presents and discusses the possible and promising research topics or directions that are based on or beyond these new methods. These topics or directions are heuristic for the researchers who are going to do further research in this area.

We start with Chapter 1 to introduce the background knowledge about computational approaches of protein function prediction in order to give an overview of this research area and to help provide a better context and understanding of the succeeding chapters. The prevalent function annotation schemes FunCat (Ruepp et al., 2004) and GO (GO Consortium, 2016) are introduced first in this chapter and are the base of function prediction. The current representative approaches and corresponding methods of function prediction, as well as their advantages, disadvantages and limitations, are then briefly reviewed and discussed. These approaches include those that are based on amino acid sequences, protein structures, genome sequences, phylogenetic data, microarray expression data, protein interaction networks, biomedical literature and the combination of multiple data types. The new approaches and methods introduced in this book are based on protein interaction networks. The motivations and purposes of this book are also presented in this chapter.

In Chapter 2, we start to discuss the innovative approaches and methods of function prediction which are the focuses of this book. Since the protein interaction data is the base of function prediction and its quality has a significant impact on prediction quality, Chapter 2 is dedicated to the new methods that assess the semantic reliability of protein interactions in a data set. After briefly introducing some representative methods of interaction reliability assessment, this chapter presents a new method for assessing the semantic reliability of interactions. This method properly incorporates topological information of an interaction network with the semantic information of proteins. This new interaction reliability assessment method is also adopted to remove unreliable interactions from the original data set to improve the data set quality semantically. Meanwhile, the commonly used approaches for evaluating the reliability assessment are also presented in this chapter.

Chapter 3 introduces two new function prediction methods that use the clustering approach. Some new protein similarities are defined accordingly. The first method excludes the target unannotated protein from clustering operations and implements the layered function prediction by selecting layered feature functions from the generated clusters. The second method predicts protein functions by dynamically tracing function occurrences across all clusters that are generated from the recursive clustering operations and are relevant to the target unannotated protein.

Iterative approaches that dynamically predict functions are introduced in Chapter 4 with three new prediction methods. New function similarities and semantic protein similarities are defined accordingly under the FunCat and GO annotation schemes. The first method, named local iterative method, iteratively makes use of local semantic information within the neighbours of the target unannotated protein to predict functions. The second method, named semi-local iterative method, exploits the local semantic information and the global importance of functions to iteratively predict functions. The third method, named global iterative method, establishes a framework within which all unannotated proteins are involved in prediction processes iteratively for function prediction. The iterative prediction approaches more reasonably reflect the nature of protein functional interactions in the prediction processes.

Instead of individually considering the contribution of proteins and their functions to the final prediction results, the functional aggregation approach introduced in Chapter 5 integrates the correlations of proteins, as well as the correlations of their functions, into function prediction. The functional aggregation is implemented by using the Choquet–Integral technique in fuzzy theory. A new semantic functional similarity of protein is defined as well in this chapter to ensure that this technique is successfully applied to function prediction.

Chapter 6 presents an innovative function prediction approach that uses a probability method to iteratively search for a proper prediction domain and predict functions accordingly. This new approach enables the prediction to achieve an optimal balance between the quality of prediction domain and the quality of prediction results and also to reduce the impact of noise information on the prediction results.

Lastly, Chapter 7 introduces a new concept and metric, protein functional connectivity, to measure the functional importance of a protein within a protein interaction data set and a new iterative prediction method that incorporates protein functional connectivity into prediction process to improve the effectiveness of prediction algorithms.

Generally, protein interaction data sets with the corresponding network model provide a flexible platform on which various techniques and information can be integrated and applied to develop new methods for protein function prediction. The approaches and methods presented in this book emphasize the semantic, correlative and dynamic features of computational prediction of protein functions. These features, as well as the deep discussions about them, make the approaches in this book unique and innovative compared to other existing approaches, and they open a gate for the researchers to do further research in this challenging and exciting area.

8.2 FUTURE DIRECTIONS

Protein function prediction has gone through the stages of using individual data sources for prediction to the stage that incorporates various data sources and techniques to achieve this target. Generally, computational protein function prediction relies on two bases: data sources and prediction models/methods as depicted in Fig. 8.1. Because no methods are widely accepted as the standard for computational protein function prediction and each existing method has its own limitations, the trend of this research area is to combine available data sources and techniques to more effectively predict functions. The combination is in terms of data source integration, model/method integration or both. This trend also applies to the research that is based on protein interactions.

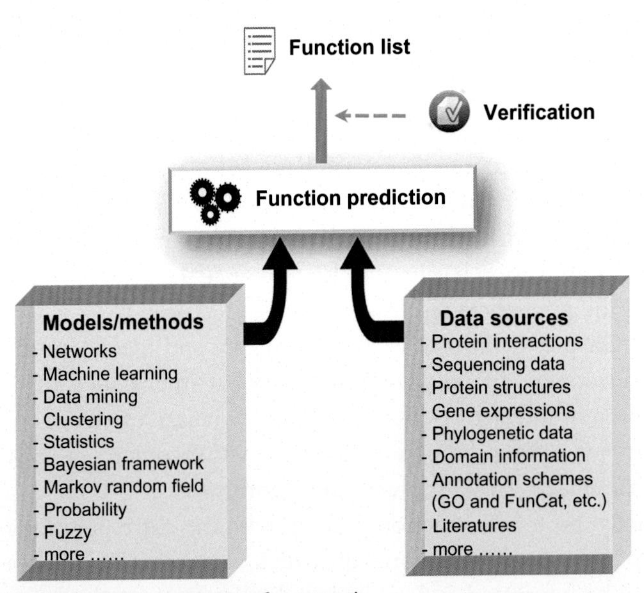

Fig. 8.1 Protein function prediction framework.

Although protein interaction is one of the available sources used for function prediction, the network model of protein interaction is a flexible platform based on which various techniques and information can be brought together for various purposes, including protein function prediction. In addition, more and more research is now focusing on incorporating semantic information into function prediction to make the results more reasonably reflect the nature of functional relationships between proteins. However, integration of different data sources for function prediction still faces many challenges. One issue of data source integration is the integrated data format or structure, since different data sources, such as gene expression data, protein interaction data and gene sequencing data, have different formats that convey information from different domains. Another issue is about the quality and reliability of data sets generated from the integration of different data resources, as the data resources usually contain considerable noise information because of errors and limitations in experiments and *silico* methods. In other words, effective data resource processing and integration is one of the research directions in this area.

In addition to data source integration, another possible and challenging research direction is the integration of various function prediction methods, as the current methods are unable to reveal and make use of all available biological information (explicit or hidden) in function prediction. The integration could be implemented by developing a middle-ware between the user and a group of available prediction methods, based on which a prediction system could be established to select the most suitable methods for function prediction based on the features of data sets, and use other available prediction methods to verify and produce optimal prediction results.

Current semantic similarities between proteins are mainly based on the structure and function definitions in annotation schemes such as the GO and FunCat. Along with data source and prediction method integrations, the semantic similarity could be defined beyond the annotation schemes by incorporating other information from various data sources to more effectively support the new developed prediction models and methods.

Current protein interaction-based prediction methods select candidate functions of an unannotated protein from the available functions in a protein interaction data set. However, many investigations have shown that the functions of a protein might not appear in its neighbour proteins or even be covered by the function set of the protein interaction data set. This phenomena indicates that only the interaction network could not fully reveal the actual biological and functional relationships among proteins. This is an unavoidable limitation of the current prediction methods. A possible way that could be taken to address this issue is to extend the function prediction domain by bringing the functions of an ontology, such as GO, into the current functional domain. This extension could be done by defining a new function similarity or using the existing function similarity to select the functions of an ontology that are similar to but not the functions of the current domain. These new selected functions from the ontology would then be added into the current function domain to form a *function network* over the protein interaction network. Based on these two networks, some new (semantic) similarities for functions and proteins could be defined and new prediction methods could be developed.

Another issue in function prediction is to keep a balance between using a mathematical model-based prediction approach and using a non-model–based approach. Mathematical model-based approaches usually predict functions based on the assumption that the relationships between proteins or protein functions follow some mathematical rules or regulations and can be mathematically modelled. Because the details and mechanisms of many biological processes are still unclear/unknown at the moment, simply harnessing the biological processes or relationships to some predefined mathematical models might result in the prediction process being unable to reflect the real relationships among proteins and functions and, in turn, the prediction results being distorted. Non-model–based approaches might be flexible because they are based on the observation of data features and try to utilize innovative ways to reflect the real relationships among proteins and functions in prediction. However, this approach might be unable to precisely describe some existing facts or behaviours. A proper combination of these two approaches at different stages of function prediction could be a possible way to greatly improve the function prediction quality.

Research on protein interaction networks could be extended beyond function prediction for other purposes, such as identifying those proteins from their interaction networks that play important roles in controlling other proteins or information transmission among proteins. This kind of research has promising applications, such as providing helpful guidance to pharmaceutical industries for designing and developing new drugs to treat cancers. Along with this research, there could be another research direction that predicts controlling directions between proteins in their interaction networks. This research could more precisely reveal the controlling relationship between proteins by adding directions into the network. Furthermore, this kind of research could be combined with the research on gene regulation networks to get more applicable results, because proteins are the products of genes, and it is obvious that any results about genes could be applied to proteins, and vice versa.

The approaches and methods presented in this book are based on the network model. Although they are for protein function prediction, the ideas and methodologies of these approaches and methods could also be adapted and applied to other research areas that rely on the network model, such as web community analysis (Zhang et al., 2006) and web service personalization (Eirinaki and Vazirgiannis, 2003), social media data analysis and customer preference and behaviour prediction (Asur and Huberman, 2010), mobile application personalization (Skillen et al., 2012) and more. In a similar way, new research approaches, methods and results from other areas could also inject fresh blood into the research of protein function prediction and other related bioinformatics research areas.

REFERENCES

Asur, S., Huberman, B.A., 2010. Predicting the future with social media. In: IEEE/WIC/ACM International Conference on Web Intelligence and Intelligent Agent Technology-vol. 1, pp. 492–499.

Eirinaki, M., Vazirgiannis, M., 2003. Web mining for web personalization. ACM Trans. Internet Technol. 3 (1), 1–27.

GO Consortium, 2016. Ontology structure. http://geneontology.org/page/ontology-structure (accessed 11.03.16).

Misteli, T., 2001. Protein dynamics: implications for nuclear architecture and gene expression. Science 291, 843–847.

Ruepp, A., Zollner, A., Maier, D., Albermann, K., Hani, J., Mokrejs, M., Tetko, I., Güldener, U., Mannhaupt, G., Münsterkötter, M., 2004. The FunCat, a functional annotation scheme for systematic classification of proteins from whole genomes. Nucleic Acids Res. 32 (18), 5539–5545.

Skillen, K.-L., Chen, L., Nugent, C.D., Donnelly, M.P., Burns, W., Solheim, I., 2012. Ontological user profile modeling for context-aware application personalization. Ubiquit. Comput. Ambient Intell. 7656, 261–268.

Zhang, Y., Yu, J.X., Hou, J., 2006. Web Communities, Analysis and Construction. Springer-Verlag, Berlin.

INDEX

Note: Page numbers followed by *f* indicate figures.

Printed in the United States
By Bookmasters